Fractured Porous Media

PIERRE M. ADLER,
JEAN-FRANÇOIS THOVERT,
VALERI V. MOURZENKO

Great Clarendon Street, Oxford, OX2 6DP,
United Kingdom

Oxford University Press is a department of the University of Oxford.
It furthers the University's objective of excellence in research, scholarship,
and education by publishing worldwide. Oxford is a registered trade mark of
Oxford University Press in the UK and in certain other countries

© Pierre M. Adler, Jean-François Thovert, Valeri V. Mourzenko 2013

The moral rights of the authors have been asserted

First Edition published in 2013

Impression: 1

All rights reserved. No part of this publication may be reproduced, stored in
a retrieval system, or transmitted, in any form or by any means, without the
prior permission in writing of Oxford University Press, or as expressly permitted
by law, by licence or under terms agreed with the appropriate reprographics
rights organization. Enquiries concerning reproduction outside the scope of the
above should be sent to the Rights Department, Oxford University Press, at the
address above

You must not circulate this work in any other form
and you must impose this same condition on any acquirer

British Library Cataloguing in Publication Data

Data available

Library of Congress Cataloging in Publication Data
Library of Congress Control Number: 2012944650

ISBN 978-0-19-966651-5

Printed and bound by
CPI Group (UK) Ltd, Croydon, CR0 4YY

Links to third party websites are provided by Oxford in good faith and
for information only. Oxford disclaims any responsibility for the materials
contained in any third party website referenced in this work.

Preface

"And further, by these, my son, be admonished:
of making many books *there is* no end;
and much study *is* a weariness of the flesh."

Proverbs, 12:12

(Authorized King James Version of the Bible)

Many people may question the reason why such a scientific book needs to be written—we, the authors, have concluded that it is probably the result of a real awareness that progress has been made over the years in a given field and that work has been done. In the last ten years, we have published more than 70 papers devoted to fractured media in one way or another. Therefore, an overall outlook emerges after such constant and long-lasting efforts.

Another important fact is that the content of this book has been taught to students many times in a wide range of countries. Generally speaking, the courses were greatly appreciated and the students claimed that they had learnt something! These comments are, of course, rewarding for the teachers of the subjects.

Repeatedly, we are reminded of the advice given by Vladimir Entov when he visited Paris for the last time. On that occasion, he criticized our papers, on the grounds that most of them were difficult to read, and recommended that we write a book which would make our studies more easily accessible.

Inevitably, we were confronted with these ideas. From the beginning, we modified a particular teaching course when certain information or developments were not immediately understood by the students; moreover, errors, confusing notations or hand-waving arguments were progressively eliminated. In time, the courses in question and the accompanying slides were enhanced to such a degree that only minor changes were required. It then appeared that the course was ready for book form.

Early in the spring of 2011, our group made some significant progress and we became convinced that we could produce a book which would be different from our previous one and which would provide a unified viewpoint of the subject.

If, after reflection, the task of writing a book is more challenging than writing a paper for a journal, it necessitates a lot of additional work even when the necessary material is available and has been published in some form. It is essential to remain as consistent and clear as possible and not to copy parts of one's own papers.

Fig. 0.1 "In my walls" by Souâd Aliane. Reprinted with permission of the artist.

Luckily, our fascination for the complexity of real fractured media and the utility of comprehending the way fluid flows in such structures were powerful motivations. Furthermore, artificial media can possess fracture-like elements. An attractive example is provided by the painting displayed in Fig. 0.1.

Once the writing of this book had begun, it occupied most of our time, including holidays—hiking without a camera became impossible—what if we came across an interesting fractured outcrop and the opportunity for a perfect picture?

Various persons collaborated with our team by contributing to this book in different ways. Among them were a number of students and post doc fellows including Igor I. Bogdanov who began permeability calculations of fractured porous media. We are also grateful to Michèle Vignes-Adler, Pierre Genthon and Ravid Rosenzweig for reading and commenting on earlier versions of the book. Gillian McAnulty-Debrabander, once again, did her utmost to polish up our English. Alice Thovert revisited many of the illustrations.

Every preface concludes with a personal message and this one is no exception. Without the affectionate and attentive support of our nearest and dearest, none of this work would have been possible and we wish to dedicate this monograph to them.

4 April 2012

Contents

1 Introduction   1
   1.1 General   1
   1.2 Description and terminology   2
   1.3 The concept of permeability   4
   1.4 Objectives and organization of this book   6
   Exercises   8

2 The geometry of a single fracture   9
   2.1 Introduction   9
   2.2 Analysis of a fracture   10
      2.2.1 The three statistical characteristics of a fracture   10
      2.2.2 The two major autocorrelation functions   12
      2.2.3 Summary   14
   2.3 Generation of random fractures   14
      2.3.1 Generation of correlated random fields   14
      2.3.2 Generation of a random fracture   16
   2.4 Geometrical properties   16
      2.4.1 Contact zones   16
      2.4.2 Spatially periodic media   18
      2.4.3 Connected and percolating components   19
   2.5 The concept of percolation   19
      2.5.1 Percolation, cluster and percolation threshold   20
      2.5.2 Power laws   21
      2.5.3 Self similarity   22
      2.5.4 Finite size effects   22
      2.5.5 The percolation thresholds for fractures   23
   2.6 Extensions   24
      2.6.1 Generation of correlated fields by Fourier transforms   24
      2.6.2 Self-affine surfaces   25
   Exercises   29

3 The geometry of fracture networks   30
   3.1 Introduction   30
   3.2 Classical analysis of a fracture network   32
      3.2.1 Analysis of traces on plane surfaces   32
      3.2.2 An example of three-dimensional field data   34
   3.3 Generation of a fracture network   35
      3.3.1 Complete analysis   35
      3.3.2 Deterministic models   36

		3.3.3	Random models	37
		3.3.4	Use of a library of data	39
		3.3.5	Some concluding remarks	40
	3.4	Statistical geometrical properties of fracture networks		40
		3.4.1	A unifying concept: the excluded volume	41
		3.4.2	The dimensionless density ρ'	43
		3.4.3	The percolation threshold of identical convex fractures	43
		3.4.4	Solid blocks	46
		3.4.5	Concluding remarks	47
	3.5	Estimation of the dimensionless density from line data		48
	3.6	Estimation of the dimensionless density from surface data		50
	3.7	Stereological relations for convex fractures		51
		3.7.1	The direct stereological relations	52
		3.7.2	The inverse stereological relations	53
		3.7.3	Consistency relations	53
	3.8	Extensions		54
		3.8.1	Percolation of fracture networks with power law size distributions	54
		3.8.2	Stereological relations for anisotropic networks of convex fractures	57
		3.8.3	The percolation threshold of anisotropic networks of convex fractures	59
		3.8.4	Stereological relations for anisotropic and heterogeneous networks of convex fractures	59
		3.8.5	Networks of heterogeneous fractures	61
	Exercises			65
4	**Transport in a single fracture**			**66**
	4.1	Introduction		66
	4.2	Flow of a Newtonian fluid		66
		4.2.1	Rigorous approach	66
		4.2.2	Analytical approximations	70
		4.2.3	Estimations of the fracture transmissivity σ	72
		4.2.4	Comparison between Reynolds and Stokes transmissivities	73
	4.3	Diffusion of a passive solute		74
		4.3.1	A rigorous approach	74
		4.3.2	The Reynolds approximation	75
		4.3.3	Estimations of the fracture conductivity Λ	76
		4.3.4	Comparison between Reynolds and Laplace conductivities	77
	4.4	Dispersion of a passive solute		78
	4.5	Extensions		81
		4.5.1	Plane channels with wavy walls	81
		4.5.2	Extension to self-affine fractures	82
	Exercises			85

5 Transport in fracture networks — 86
- 5.1 Introduction — 86
- 5.2 General — 87
  - 5.2.1 Flow equations — 87
  - 5.2.2 Permeability K_n of a fracture network — 87
  - 5.2.3 Flow through networks made of infinite fractures — 89
- 5.3 Numerical methodology — 91
  - 5.3.1 Meshing — 91
  - 5.3.2 Discretization of the equations and resolution — 92
  - 5.3.3 Network permeability — 93
  - 5.3.4 An elementary example — 93
- 5.4 I^{2}OUD fracture networks — 94
  - 5.4.1 Permeability — 94
  - 5.4.2 Properties of the velocity field — 98
- 5.5 Extensions — 102
  - 5.5.1 Fracture networks with power-law size distributions — 102
  - 5.5.2 Anisotropic fracture networks — 104
  - 5.5.3 Networks of heterogeneous fractures — 106
- Exercises — 108

6 Transport in a fractured porous medium — 109
- 6.1 Introduction — 109
- 6.2 General — 109
  - 6.2.1 Flow equations — 109
  - 6.2.2 Permeability K_{eff} of a fractured porous medium — 111
  - 6.2.3 A simple approximation for K_{eff} — 112
- 6.3 Numerical methodology — 113
  - 6.3.1 Meshing — 113
  - 6.3.2 Discretization of the equations and resolution — 116
  - 6.3.3 The permeability of the fractured porous medium — 117
  - 6.3.4 An elementary example — 117
- 6.4 Fractured porous media with I^{2}OUD fractures — 118
  - 6.4.1 Importance of the existence of the percolation threshold on individual samples — 118
  - 6.4.2 Influence of the fracture shapes — 121
- 6.5 Extensions — 121
  - 6.5.1 Fractured porous media with power-law distribution of fracture sizes — 121
  - 6.5.2 Slightly compressible flows and application to pressure drawdown well tests — 125
- Exercises — 128

7 Two-phase flow through fractured porous media — 129
- 7.1 Introduction — 129
- 7.2 Local equations on the Darcy scale — 129

		7.2.1	Conservation equations	129
		7.2.2	Constitutive equations	131
	7.3	Numerical approach		133
		7.3.1	Spatial discretization	133
		7.3.2	Resolution of the equations	134
	7.4	Regular fracture networks		134
		7.4.1	Multiple families of parallel plane fractures	134
		7.4.2	Sugar-box reservoir	135
	7.5	Isotropic and homogeneous fractured porous media		138
		7.5.1	Transients from various initial states	138
		7.5.2	Steady state macroscopic properties	139
		7.5.3	Influence of the parameters	143
	7.6	Comparison with a capillary dominated model		143
	Exercises			145
8	**Concluding remarks**			**146**
	8.1	Introduction		146
	8.2	Numerical		146
		8.2.1	Two-dimensional mesher	146
		8.2.2	Three-dimensional mesher	148
		8.2.3	Single-phase flow	153
		8.2.4	Two-phase flow	155
		8.2.5	Conclusions	156
	8.3	Other phenomena		156
	8.4	Where do we stand?		158
Notation				**159**
References				**167**
Index				**173**

Introduction

1.1 General

The general objective of this book is to estimate the macroscopic properties of fractures, fracture networks and fractured porous media from easily measurable quantities. Attention is focused on geological media where rocks such as sandstones, carbonates and granites are necessarily fractured at various scales by the slow but constant motion of continental masses. This book is situated between three different disciplines. First, geology and geophysics provide most of the data and most applications; a characteristic feature is that one never has a complete knowledge of the studied objects—such as an oil reservoir—in contrast with a laboratory experiment where every quantity can be measured. Second, engineering develops the main tools of analysis such as flow and permeability calculations. Third, statistical physics plays a major role in concepts such as the excluded volume, continuum percolation and power laws. In view of this interdisciplinary character, the general results presented in this book may have unexpected applications in many different domains.

This book is based on courses which have been taught in several countries at Master and Ph.D. levels in universities, research centers and at conferences. It should provide, in a compact form, all the necessary tools to achieve the general objective. The mathematical level in this book has been kept as low as possible. The interested reader can always go further thank to the references that are provided and in which the mathematical level is not as restricted. Also, we have preserved the colloquial aspect of a course where one tries to explain abstract concepts in simple terms.

For pedagogical purposes, the bulk of this book deals with the isotropically oriented and uniformly distributed networks of fractures; in addition, the fractures are supposed to be *monodisperse* which means that they are all of the same size. But, most real fracture networks are obviously different. Therefore, Chapters 2–6 each end with a section called "Extensions".

These Extensions are devoted to more complex and more recent developments. They should be skipped by the beginner since they are very hard to understand without reading the original papers. Nevertheless, they can be used in two ways. First, they can be browsed very rapidly in order to know that some progress has been made in a particular area. Second, the sequence of formulae which is provided, is, in most cases,

1.1	General	1
1.2	Description and terminology	2
1.3	The concept of permeability	4
1.4	Objectives and organization of this book	6
Exercises		**8**

Fig. 1.1 A view of a mountain in the south of France in (a) (Photography by Ph. Carri/FOL48). The magnifying glass shows us what is in (b) with two items; (1) is an apparently solid medium while (2) corresponds to discontinuities or fractures. When magnified, the apparently solid medium in (b) may look like (c); it is actually a porous medium; the solid phase is white while the pores whose characteristic size is ℓ_p, are black.

complete; therefore, an experienced reader can use these sections just as with the classical book by Gradshteyn and Ryzhik (1965); all the material is available for a quick estimate of a macroscopic property of a fractured porous medium with specific properties.

An important aspect of this book is that short exercises propose systematic applications of the major concepts and results right after their introduction. Usually, orders of magnitude are also calculated. Some exercises are incomplete in the sense that one has to make assumptions, as is the case in a real practical problem where not all the necessary data are available. These exercises are very useful, since they oblige the attendees to play an active role, and we insist that they actually make the numerical applications. Their resolution stimulates general discussion, questions, remarks and criticisms which greatly benefit the atmosphere of the course. Finally, the exercises in the Extensions are straightforward applications of the formulae in order to help the reader to master the material.

A final point should be made about the references given in this book. A relatively complete account of the literature prior to 1999 is provided by Adler and Thovert (1999); since the present book is a monograph, it does not provide an updated survey. However, over the last ten years, progress has been made by others and it cannot be ignored; several books have been published and they should be briefly mentioned here. Collective contributions have been gathered by Faybishenko *et al.* (2005) and Dietrich *et al.* (2005). The latter is written from an engineering perspective and is quite complementary to ours. This is also valid for the books by Nelson (2001) and Singhal and Gupta (2010).

The most recent is the second edition of the classical book by Sahimi (2011). It contains several chapters devoted to fractures, fracture networks and fractured porous media with a similar point of view to ours.

1.2 Description and terminology

It may be convenient to start with real examples which can be seen almost everywhere, when one leaves large cities. We can look first at Fig. 1.1a which shows a partly bare mountain from a distance; the bare part looks like a steep cliff. If one uses binoculars, there is something which looks like (b), an almost vertical wall with two distinct components. There are massive pieces of rocks which apparently contain no void space. These pieces are separated by discontinuities which are more or less linear and which are called *fractures*. Suppose that one now uses a magnifying glass to look more closely at a given piece of the solid matrix; one may then see something like what is shown in (c), namely a large number of small holes whose dimension ℓ_p is typically of a few microns. Therefore, the solid matrix is a *porous medium*.

An important remark is that the volumes of the porous medium and of the pores are much larger than the corresponding volume of the fractures. Therefore, most of the underground fluid is contained in the porous matrix.

The discontinuities that are seen in Fig. 1.1b are void spaces between two rough surfaces as detailed in Fig. 1.2a where the axes are defined. The two rough surfaces oscillate around planes parallel to the xy-plane at least locally; therefore, a fracture can be approximated by a two-dimensional void space, since the extension along the z-axis is much smaller than the extensions along the x- and y-axes. The local distance b between the two surfaces is called the local aperture. It is difficult to give a precise order of magnitude of the average value of b, but a value like 0.1 mm is reasonable for fractures whose lateral extensions are of the order of a few meters.

Therefore, the fracture aperture b is typically ten times larger than the average pore dimension ℓ_p. In Section 1.3, we shall see that the ease with which a fluid flows through a given medium is proportional to the permeability of this medium and that the permeability is roughly proportional to the square of a characteristic dimension of the medium.

A direct and important consequence of this remark is that fluids flow a hundred times more easily through fractures than through porous media. In other words, the macroscopic properties of fractured media can be drastically modified by the presence of fractures.

There might be some ambiguity between our terminology and that used by geologists. Here, a fracture refers to a discontinuity in a rock, whatever its cause may be, while for a geologist a fracture often implies a discontinuity and a displacement parallel to the discontinuity plane.

Several fractures compose a *fracture network*. A real example is displayed in Fig. 1.2 which is taken from Gonzalez-Garcia *et al.* (2000); each fracture is represented by a triangulated surface of a given grey level. It is a very important example in that it is a unique measurement of a real fracture network inside a granite cube of approximately 50 cm. The amount of work required to obtain these fractures is significant and will be detailed in Section 3.3.1. Obviously, the fractures are roughly plane and they mostly belong to two families, A and B, of different orientations.

Our general terminology should also be made more precise. When one speaks of a *fracture network*, fluid is only contained in the fractures, and the medium located in between the fractures is impervious. However, when one speaks of a *fractured porous medium*, the solid matrix located in between the fractures is porous and permeable; fluid flows inside the fractures and in the porous matrix and is transferred from one to the other according to laws which are given in Section 6.2.1.

Another term which may be misleading is *monodisperse*; a fracture network is monodisperse when all the fractures are of the same size, as already defined in Section 1.1. When the fractures are of different sizes, the network is called *polydisperse*.

(a)
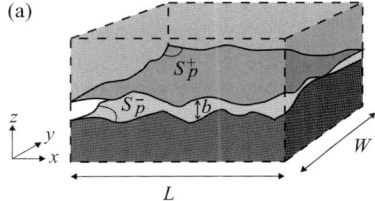
(b)

Fig. 1.2 A single fracture. (a) displays a schematized fracture of length L and width W as a void space between two surfaces S_p^+ and S_p^-. (b) shows a view of a real fracture surface (reprinted with permission from Ameen, 1995).

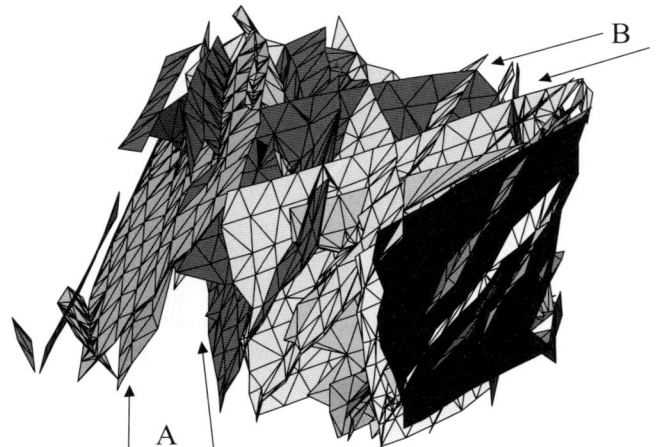

Fig. 1.3 A real fracture network in a granite block of about $(0.5)^3 \text{m}^3$. There are two major families denoted by A and B.

1.3 The concept of permeability

Consider an experiment where one wants to characterize the flow properties of a piece of material without knowing what is inside. One prepares a piece of this material and it is usual practice to take a circular cylinder (or a parallelepipedon) of section S and of length L (see Fig. 1.4). The lateral surface is made impermeable by an experimental trick, such as the gluing of an impermeable tissue around the cylinder. Then, the medium is filled with a Newtonian fluid of viscosity μ and the upstream and downstream sections are immersed in vessels at pressures p_1 and p_2 with $p_1 > p_2$. As an obvious consequence, the fluid flows from the higher pressure to the lower one, and the flow rate Q is measured. If we consider that the phenomenon is linear, we may assume that Q is proportional to the driving force $p_2 - p_1$

$$Q \propto -(p_2 - p_1) \tag{1.1a}$$

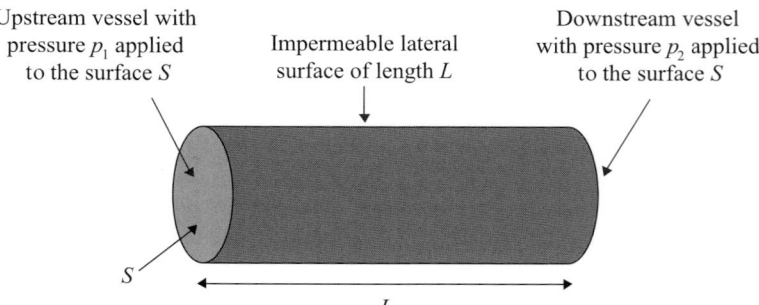

Fig. 1.4 Illustration of the concept of permeability.

The flow rate is taken to be positive when the fluid flows from 1 to 2. One may suppose that the medium is relatively homogeneous though it is useful to insist on the fact that one does not know at this point how the material is made up. If it is homogeneous, it is usual to say that the flow rate is divided by two when length L is multiplied by two for the same pressure drop $p_2 - p_1$; therefore, the quantity which matters is the pressure gradient that is approximately denoted by $\overline{\overline{\partial p/\partial x}}$

$$Q \propto -\frac{(p_2 - p_1)}{L} = -\overline{\overline{\frac{\partial p}{\partial x}}} \quad (1.1b)$$

Again because of homogeneity, one can state that the flow rate is multiplied by two when the surface S is multiplied by two for the same pressure drop $p_2 - p_1$; therefore, the quantity which matters is the seepage velocity $\overline{\overline{v}}$

$$\overline{\overline{v}} = \frac{Q}{S} \propto -\overline{\overline{\frac{\partial p}{\partial x}}} \quad (1.1c)$$

Then, if the fluid is twice as viscous, the seepage velocity is expected to be divided by two; therefore,

$$\overline{\overline{v}} \propto -\frac{1}{\mu}\overline{\overline{\frac{\partial p}{\partial x}}} \quad (1.1d)$$

The proportionality constant between the two sides of this relation is called permeability and is denoted by K; it has the dimensions of the square of length

$$\overline{\overline{v}} = -\frac{K}{\mu}\overline{\overline{\frac{\partial p}{\partial x}}} \quad (1.1e)$$

This equation was first established empirically by Darcy (1856) and it was only set on a firm theoretical basis in the 1950s. Let us give some general properties for this important quantity. First, as already stated, K is homogeneous to the square of a length. In principle, it should be expressed in m^2, but it is usually expressed in darcy (see Exercise 1.1)

$$1 \text{ darcy} \approx 1\mu\text{m}^2 = 10^{-12}\text{m}^2 \quad (1.2)$$

As pore dimensions of geological materials are of the order of a few microns, the darcy is a convenient unit.

The concept of permeability is valid in two important conditions which correspond to the same linearity condition. The first one is that the fluid should be Newtonian, i.e. the fluid stresses should be proportional to the fluid strains. The second condition is that inertial forces should be small when compared to viscous forces; this corresponds to the fact that the Reynolds number should be small; this number is usually defined as follows

$$Re = \frac{\rho_f v \ell}{\mu} \quad (1.3a)$$

where ρ_f is the fluid density, v a characteristic velocity and ℓ a characteristic scale of the pores where flow occurs. Therefore, the condition on the Reynolds number is

$$Re \ll 1 \tag{1.3b}$$

A crucial property is that K depends only on the geometry of the medium. Formally,

$$K = f(\text{geometry}) \tag{1.4}$$

This is indeed essential and explains why such great importance is devoted to geometry in this book.

Finally, one can generalize eqn 1.1e in several ways. First, this scalar relation can be made vectorial. As a rule, boldface quantities are vectors or tensors, while non-bold letters correspond to scalars. Second, K should be replaced by the tensor $\boldsymbol{K}$ when the void structure of the medium is anisotropic. Therefore, Darcy's law can be generalized as

$$\overline{\overline{\boldsymbol{v}}} = -\frac{\boldsymbol{K}}{\mu} \cdot \overline{\overline{\nabla p}} \tag{1.5}$$

where $\overline{\overline{\boldsymbol{v}}}$ is the vectorial seepage velocity. For an incompressible fluid, mass conservation implies the continuity equation

$$\nabla \cdot \overline{\overline{\boldsymbol{v}}} = 0 \tag{1.6}$$

The derivation of the basic operators is proposed in Exercise 1.2.

1.4 Objectives and organization of this book

As already stated in general terms at the beginning of Section 1.1, the objective of this book is the determination of the permeability of three structures, namely fractures, fracture networks and fractured porous media from easily measurable quantities.

The methodology which is going to be employed comprises four steps. First, since geometry plays such a crucial role, the various structures are quantitatively described.

Second, the permeability of each structure is determined. This requires going from one scale to the next larger one; the local equations on the small scale are solved and the solutions are averaged on the large scale. The average is indicated by an overbar. For instance, the local velocity in the pores of a porous medium is $\boldsymbol{v}$; the seepage velocity $\overline{\boldsymbol{v}}$ is the local velocity averaged over a large number of pores. The seepage velocity $\overline{\overline{\boldsymbol{v}}}$ on the scale L of the block of Fig. 1.4 requires another change of scale; therefore, an extra overbar is added to indicate this second change of scale. Specific numerical tools are devised in order to solve the relevant flow equations at each scale.

Third, the systematic numerical results need to be rationalized. A key quantity is the so-called *excluded volume* which was introduced in fracture studies by Balberg *et al.* (1984). When used, properties are almost independent of the fracture shape, which is a simplification of great importance.

Fourth, the permeabilities of fracture networks and of fractured porous media are estimated from easily measurable quantities.

This methodology is the red thread which marks all the parts of this book.

The book itself is organized as follows. It is divided into two parts if we exclude this Introduction and Chapter 8.

The first part deals with geometry. First, Chapter 2 addresses the random geometry of real fractures. It shows how this geometry can be characterized and reproduced numerically. Two kinds of fractures are distinguished, depending on their spatial organization, namely the Gaussian and the self-affine fractures.

Second, Chapter 3 deals with the geometry of random networks. The fractures are generally considered as plane objects such as polygons. The most important question relative to these networks is whether they percolate or not. In simple terms, can a Maxwell demon or the like go through the whole medium by walking on the fractures without any jump? The concept of percolation is detailed and applied to fracture networks in conjunction with the excluded volume. The dimensionless density is defined as the number of fractures per excluded volume and is shown to control percolation; this quantity is crucial since it plays a major role in the permeability of fracture networks and of fractured porous media as well. Then, the dimensionless density is estimated from data which can be measured along lines (wells, outcrops, ...) or on surfaces (quarries, outcrops, ...). This chapter ends with extensions such as fractures with power-law size distributions and anisotropic orientations.

The second part deals mainly with permeability under steady conditions and it occupies most of the rest of the book. Chapter 4 determines the transport properties of a single fracture. Permeability is calculated by solving the Stokes equation numerically in random fractures, generated as indicated in Chapter 2. The Reynolds approximation is also detailed and its results are systematically compared with the former exact ones. Conductivity of fractures is obtained by solving the Laplace equation and basically the same developments are made as for permeability. Some indications are given on dispersion when diffusion and convection interact to disperse a solute. Attention is focused on Gaussian fractures, though most real fractures are self-affine, but since their properties are much more difficult to obtain, they are only addressed briefly in Section 4.5.

Chapter 5 is devoted to the permeability of fracture networks. The general equations and boundary conditions which apply to networks are given. The flow inside each fracture is described by a two-dimensional Darcy equation. The Snow equation, which is valid for infinite fractures, is derived. Then, the meshing of one fracture is described when the intersections with the other fractures are taken into account. The discretization of the local equations and of the boundary conditions is made by the familiar finite volume technique. The numerical tools are first applied to isotropically and uniformly distributed networks of monodisperse fractures. The most important result is that when expressed in terms of the dimensionless density, permeability depends only very slightly on the fracture shape when it is convex. The chapter ends with a survey of the major extensions of this result to other kinds of fracture networks.

Chapter 6 extends the previous chapter to fractured porous media, and its structure is basically the same, namely a general section with the governing equations, the meshing of the solid porous matrix located in between the fractures, the discretization, the applications to isotropically and uniformly distributed networks of monodisperse fractures and the extensions. The major difficulty in these studies is the meshing of the three-dimensional solid matrix and it was quite difficult to obtain a numerical program which worked correctly and efficiently. Again the dimensionless density is shown to play an essential role in macroscopic permeability and it is used systematically to rationalize the results. Extensions are presented for networks with power-law size distributions and for slightly compressible fluids.

Chapter 7 ends the second part and provides some preliminary results on two-phase flows in fractured porous media. The classical equations on the Darcy scale are presented, as well as the van Genuchten constitutive relations. Some numerical details are given, but basically the same finite volume technique is used. The two classical examples of parallel fractures and sugar box reservoirs are briefly illustrated and discussed. Then, the general situation of two-phase flow in random networks is addressed; first, transients in a three-dimensional fractured porous medium are presented; second, some examples of upscaled permeabilities are given and discussed; systematic calculations made in the capillary dominated regime for two media with different fracture densities are well approximated by a simple model.

Finally, Chapter 8 combines some general remarks on parallel computations, recent extensions of the possibilities of the numerical codes, other situations and other phenomena which can be addressed via the same techniques.

Exercises

(1.1)   (i) Give the fundamental SI units.
(ii) Find the exact definition of one darcy.
(iii) Give an order of magnitude of the Reynolds number for water in a typical flow in a porous medium.

(1.2) Express ∇p, $\nabla \cdot \mathbf{v}$ and $\nabla \cdot (\nabla p) = \nabla^2 p$ in Cartesian coordinates.

The geometry of a single fracture

2.1 Introduction

When they consider individual fractures, geologists often use a precise terminology which describes some of the characteristics of these objects.

First, some information can be given as to the mechanical process which creates the fracture. A *fault* is defined as a fracture on which there has been an appreciable shear displacement of the material located on either side; shear means that the motion occurred parallel to the fracture plane, as illustrated in Fig. 2.1a. Since the surfaces are rough, a fault necessarily involves some crushed material which can be considered as a porous medium (see Section 6.2.1) except when it is clogged by secondary mineral precipitation due to fluid circulation inside the fault zone.

A *joint* is an open fracture (i.e. non-filled with some mineral deposit) in which there is no appreciable shear displacement, but an opening displacement normal to the fracture plane, as illustrated in Fig. 2.1b. This kind of discontinuity includes bedding joints in sedimentary rocks and cooling joints in initially hot rock masses such as lavas.

A *vein* is a mineralized fracture through which flow is no longer possible, as shown in Fig. 2.1c. Before its mineralization, the vein may or may not have undergone some shear displacement; however, some opening motion of the fracture necessarily occurred for the free space to have been created. Note that fluid circulates easily in the large fractures which are more likely to be clogged than the small ones.

Different terms are also employed, such as *fissures*, *cracks*, ..which usually designate fractures of relatively small extensions.

In this book, the generic term fracture is used and no further attention will be paid to the precise process which created the fracture.

This chapter is organized as follows. The geometrical quantities which characterize the structure of a fracture are analyzed in Section 2.2. Section 2.3 is a schematic presentation of the generation of random fractures which possess these characteristics. The resulting geometrical properties are provided in Section 2.4. The concept of percolation is briefly presented in Section 2.5. Finally, the generation of correlated fields and the properties of self-affine fields are summarized in Section 2.6.

2.1	Introduction	9
2.2	Analysis of a fracture	10
2.3	Generation of random fractures	14
2.4	Geometrical properties	16
2.5	The concept of percolation	19
2.6	Extensions	24
Exercises		29

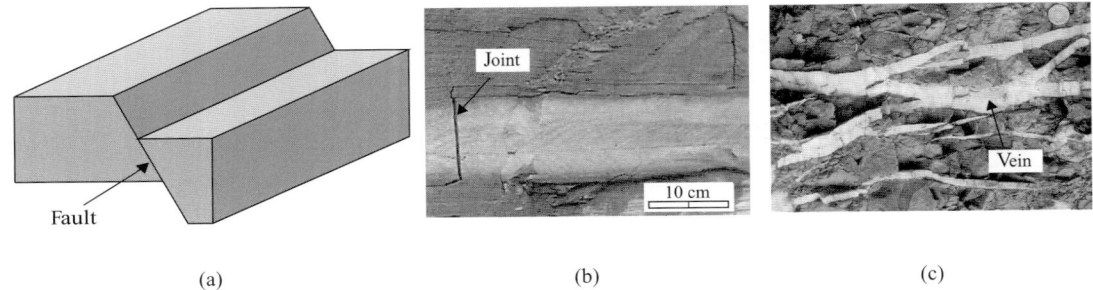

Fig. 2.1 Illustration of the geological terminology relative to fractures: (a) a fault, (b) a joint, (c) veins (the scale is indicated by a coin in the upper right corner). (b) and (c) are reprinted with permission from Ameen (1995).

2.2 Analysis of a fracture

2.2.1 The three statistical characteristics of a fracture

A fracture is limited by two surfaces which may coincide over what is called a contact area. Such a fracture is schematized in Fig. 2.2.

As shown in Fig. 1.2, the fractures are more or less plane at least over distances which are significantly larger than the typical distance between the two limiting surfaces which is called the aperture. Therefore, it is generally assumed that the two surfaces oscillate around two average planes $h_0^\pm$. The xy-plane is assumed to be parallel to these average planes. The components of the two-dimensional position vector $\boldsymbol{x}$ are equal to x and y. The two surfaces S_p^+ and S_p^- displayed in Fig. 2.2 can be represented by the relations

$$z^\pm(\boldsymbol{x}) = h_0^\pm + h^\pm(\boldsymbol{x}) \tag{2.1}$$

$h^\pm(\boldsymbol{x})$ denotes the fluctuations of the surface around the average plane. The local aperture b is first defined by

$$b(\boldsymbol{x}) = z^+(\boldsymbol{x}) - z^-(\boldsymbol{x}) \tag{2.2}$$

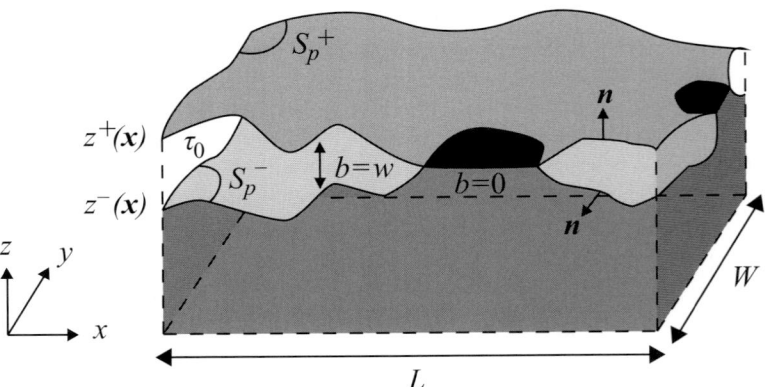

Fig. 2.2 Convention and notations for the fracture geometry. The contact areas are displayed in black.

This aperture can be equal to zero, in which case it corresponds to the contact areas which are shown in black in Fig. 2.2. However, b needs to be more precisely defined and this will be done by (2.21).

The fluctuations $h^\pm(\boldsymbol{x})$ are random in character. Let us first consider the average of these fluctuations. Two averages can be used, namely statistical and spatial averages which are denoted by brackets $<.>$ and overbars $\overline{(.)}$, respectively. For instance, the seepage velocity $\overline{\overline{v}}$ defined by (1.1c) is obtained after two volume averages, as explained in Section 1.4.

Further details regarding these two possible averages are given in Section 2.6.2. In most of this book, they are equivalent and one of them is chosen because of its convenience. For instance, experimentally it is more convenient to perform a spatial average over one fracture and that is the one which is going to be used in this section.

Since the planes $h_0^\pm$ are the average planes, the surface average of the fluctuations is equal to zero. Therefore,

$$\overline{h^\pm(\boldsymbol{x})} = 0 \qquad (2.3)$$

The distance between the two average planes is generally denoted by

$$b_m = h_o^+ - h_o^- \qquad (2.4)$$

At least, three statistical characteristics are necessary to describe the random properties of a fracture. The first one is relative to the probability densities $\varphi(h^\pm)$ of the fluctuations $h^\pm(\boldsymbol{x})$. Experimental measurements generally show that these densities are Gaussian with identical variance

$$\varphi(h^\pm) = \frac{1}{\sqrt{2\pi}\sigma_h} \exp\left[-\frac{h^{\pm 2}}{2\sigma_h^2}\right], \qquad \sigma_h = \sigma_{h^+} = \sigma_{h^-} \qquad (2.5a)$$

σ_h is also called the roughness of the surfaces.

The second statistical characteristic is relative to the organization of each surface. A common way to characterize these organizations is to introduce an autocorrelation function

$$C_{h^\pm}(\boldsymbol{u}) = \overline{h^\pm(\boldsymbol{x})h^\pm(\boldsymbol{x}+\boldsymbol{u})} \qquad (2.5b)$$

where $\boldsymbol{u}$ is the lag. The average is performed over the space variable $\boldsymbol{x}$. For instance, if the average is taken over a line of length L, one has

$$C_{h^\pm}(u) = \frac{1}{L}\int_0^L h^\pm(x)h^\pm(x+u)\,\mathrm{d}x \qquad (2.5c)$$

For isotropic surfaces, $C_{h^\pm}$ only depends on the norm u of $\boldsymbol{u}$. Generally speaking, the correlation functions of the two surfaces are equal, i.e. $C_{h^+}(u) = C_{h^-}(u)$.

The autocorrelation function has two important properties. For a zero lag, the autocorrelation function is equal to the square of the roughness, i.e. the variance of $h^\pm(\boldsymbol{x})$

$$C_{h^\pm}(u=0) = \sigma_h^2 \qquad (2.6a)$$

For large values of u, the surface fluctuations are independent and because of (2.3),
$$C_{h\pm}(u = \infty) = 0 \tag{2.6b}$$

The third statistical characteristic is the interrelation between the two surfaces. This interrelation is quantified by the intercorrelation I_h between the two surfaces
$$I_h = \overline{h^+(\boldsymbol{x})\,h^-(\boldsymbol{x})} \tag{2.7a}$$

The intercorrelation coefficient is defined as (see Exercise 2.1)
$$\theta_I = \frac{I_h}{\sigma_h^2} \tag{2.7b}$$

2.2.2 The two major autocorrelation functions

There are two major classes of autocorrelation functions $C_h(\boldsymbol{u})$, namely the Gaussian and the self-affine autocorrelations. From now on, the statistical properties of the upper and lower surfaces are identical and the superscript $\pm$ is omitted.

The Gaussian autocorrelation function for an isotropic fracture can be expressed as
$$C_h(u) = \sigma_h^2 \exp\left[-\left(\frac{u}{\ell_c}\right)^2\right] \tag{2.8}$$

This autocorrelation is characterized by a single correlation length scale ℓ_c, and the fracture surfaces with such an autocorrelation look like old mountains. An example is given in Fig. 2.3. ℓ_c corresponds roughly to the distance between two summits or two valleys. The variations in the "landscape" with ℓ_c are illustrated by the two overall views in Figs 2.3b and c, and by the two cuts in e and f. It is easy to generate anisotropic Gaussian fractures since (2.8) can be generalized as
$$C_h(\boldsymbol{u}) = \sigma_h^2 \exp\left[-\left(\frac{u}{\ell_{c1}}\right)^2 - \left(\frac{v}{\ell_{c2}}\right)^2\right] \tag{2.9}$$

where u and v are now the two components of $\boldsymbol{u}$. The two characteristic length scales ℓ_{c1} and ℓ_{c2} correspond to the two axes. An example of an anisotropic surface is displayed in Fig. 2.3d.

The Gaussian surfaces are statistically homogeneous, or in other words they are statistically invariant by arbitrary translations. In simple terms, this means that the average properties are independent of the position. For instance, the surface roughness and the correlation lengths are constant in space.

Very different surfaces can be generated by a simple generalization of (2.8).
$$C_h(u) = \sigma_h^2 \exp\left[-\left(\frac{u}{\ell_c}\right)^{2H}\right] \qquad 0 \leq H \leq 1 \tag{2.10}$$

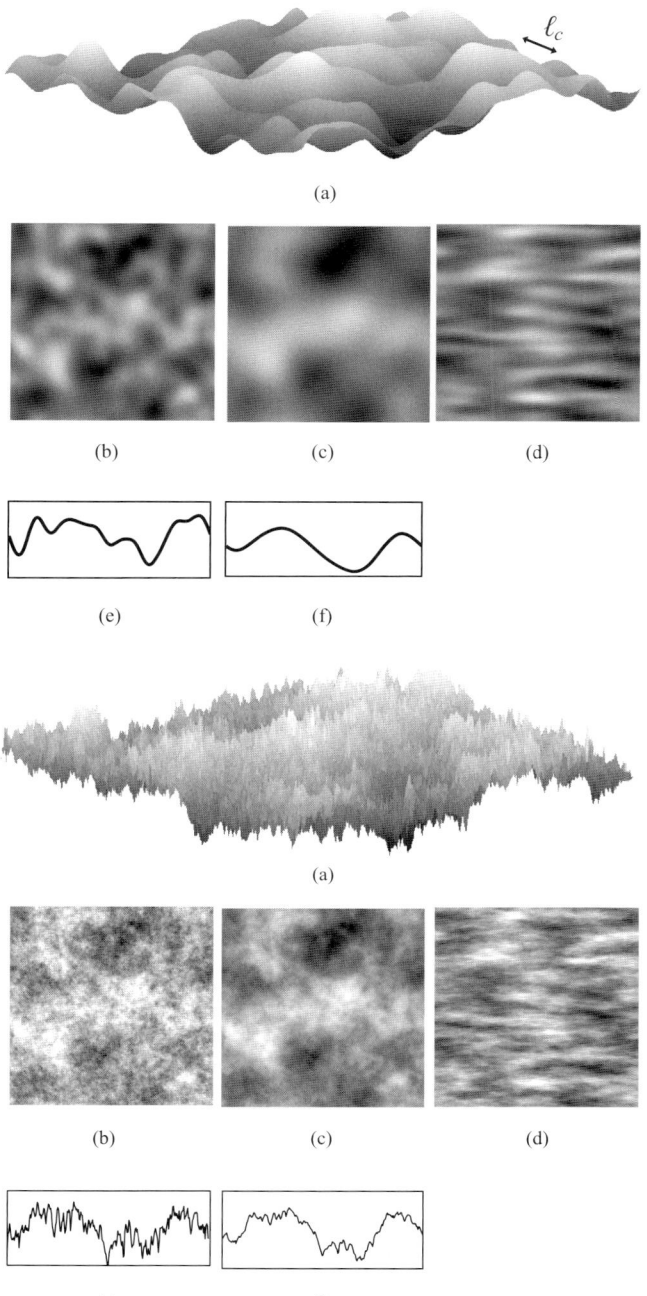

Fig. 2.3 Standard Gaussian fields with the Gaussian correlation function (2.8). (a) Three-dimensional view of a random surface with a Gaussian correlation; (b), (c) and (d) are overhead views of random surfaces with a Gaussian correlation; the size of the images is 256×256; (e) and (f) are transversal cuts along the lower side of (b) and (c), respectively. Data are for: (a) and (b) ($\ell_c = 64/\pi$), (c) ($\ell_c = 128/\pi$), (d) ($\ell_{c1} = 128/\pi, \ell_{c2} = 32/\pi$).

Fig. 2.4 Self-affine surfaces: (a) three-dimensional view of a random surface with a self-affine correlation; (b) and (c) are examples of isotropic self-affine surfaces generated by the Fourier transform technique; (d) anisotropic self-affine surface; the curves (e) and (f) correspond to cuts made along the lower side of (b) and (c), respectively. Data are for: (b) and (e) ($H = 0.25$), (c), (d) and (f) ($H = 0.75$).

These surfaces are called *self-affine*. The previous relation was suggested by Sinha *et al.* (1988). H is the Hurst exponent. $H = 1$ corresponds to regular surfaces with Gaussian correlations, while the surfaces are self-affine for $H<1$. The expression (2.10) implicitly introduces an upper cut-off length ℓ_c for self-affinity, above which all correlations disappear. Accordingly, the variance σ_h^2 is finite and the surfaces are statistically homogeneous at scales much larger than ℓ_c.

More information on these self-affine surfaces is given in Section 2.6.2.

Although the generalization (2.10) of (2.8) is a priori insignificant, the corresponding surfaces are very different, since they look like young spiky mountains as illustrated in Fig. 2.4a. Other representations are displayed in (b) and (c) for two values of the Hurst exponent. It is interesting to consider the cross sections of these surfaces by a plane which are illustrated in (e) and (f); when H is close to 1, the line which corresponds to the cross section is "linear" (and close to the relatively smooth cuts obtained for Gaussian surfaces in Fig. 2.3), while it gets "thicker" when H is small and close to 0. This is due to the fact that H is a decreasing function of the fractal dimension of the surface. Fractals are only briefly alluded to in Subsection 2.5.3 with adequate references.

As seen previously, anisotropic self-affine surfaces can be generated by the following correlation function

$$C_h(\boldsymbol{u}) = \sigma_h^2 \exp\left[-\left\{\left(\frac{u}{\ell_{c1}}\right)^2 + \left(\frac{v}{\ell_{c2}}\right)^2\right\}^H\right] \qquad (2.11)$$

Such an anisotropic surface is represented in Fig. 2.4d.

2.2.3 Summary

The information contained in the two previous subsections can be summarized as follows:

A fracture is characterized by a series of quantities. The first two are its average aperture b_m and its roughness σ_h. The nature of the surface needs to be defined; it can be Gaussian or self-affine; the length scale ℓ_c appears in both cases while the Hurst exponent H is only necessary in the latter. The intercorrelation coefficient θ_I is also needed.

Alternatively, one can introduce dimensionless quantities and it is usual to consider σ_h as the unit length. Therefore, the first two quantities are σ_h and b_m/σ_h. The nature of the surface should be defined as before; ℓ_c/σ_h and possibly H are needed as well as the intercorrelation coefficient θ_I.

At this point, it is important to add that Gentier (1986) and Brown et al. (1986) showed that typical experimental values belong to the intervals

$$0 < \frac{b_m}{\sigma_h} < 2.6, \quad 1 < \frac{\ell_c}{\sigma_h} < 7 \qquad (2.12)$$

2.3 Generation of random fractures

2.3.1 Generation of correlated random fields

In order to illustrate the main ideas, consider a one-dimensional random field $X(i)$ ($i = 1, 2, \cdots, \infty$). $X(i)$ is supposed to be a Gaussian variable of zero mean and of variance 1; $X(i)$ is independent of $X(j)$ when $i \neq j$. In other words,

$$<X(i)> = 0, \quad <X^2(i)> = 1, \quad <X(i)X(j)> = 0 \text{ when } i \neq j \quad (2.13)$$

Such a field can be easily generated on any modern computer.

The basic way to derive a correlated field from an uncorrelated one is to make a linear combination of the independent values. For instance, consider the random field $Y(i)$ defined as

$$Y(i) = \sum_{m=0}^{L_c} a(m) X(i+m) \quad (2.14)$$

where L_c is a positive integer which plays more or less the role of a correlation length. $[a(m); m = 1, \ldots, L_c]$ are coefficients which are a priori unknown. Since $Y(i)$ is a linear combination of Gaussian variables, it is a Gaussian variable whose mean is obviously zero. Its variance can be easily determined and imposed as equal to 1 (see Exercise 2.2)

$$<Y^2(i)> = \sum_{m=0}^{L_c} a^2(m) = 1 \quad (2.15)$$

Therefore, the field $Y(i)$ also has a variance which is equal to 1.

The autocorrelation function of the field Y can be expressed in a general way as

$$<Y(i)Y(j)> = \left\langle \sum_{m=0}^{L_c} a(m) X(i+m) \sum_{n=0}^{L_c} a(n) X(j+n) \right\rangle$$
$$= \sum_{m=0}^{L_c} \sum_{n=0}^{L_c} a(m) a(n) \langle X(i+m) X(j+n) \rangle, \quad i<j \quad (2.16)$$

where it is assumed without any loss in generality that i is strictly smaller than j. Because of (2.13), the only terms which remain verify

$$i + m = j + n \quad (2.17a)$$

This necessitates that

$$j - i \leq L_c \quad (2.17b)$$

Therefore,

$$<Y(i)Y(j)> = \sum_{m=j-i}^{L_c} a(m) a(m+i-j) \text{ for } j-i \leq L_c \quad (2.18)$$

One can express this in terms of the linear autocorrelation function (2.5c) with $k = j - i$

$$C_Y(k) = \sum_{m=k}^{L_c} a(m) a(m-k) \text{ for } k \leq L_c \quad (2.19)$$

These simple calculations demonstrate a very important property. One can build correlated random fields Y by using linear combinations of uncorrelated random fields X. The autocorrelation function C_Y is a quadratic function of the coefficients a of the linear combinations. Moreover, in principle at least, the coefficients a can be derived from C_Y by solving the set of non-linear equations (2.19); C_Y can possibly be given by (2.8).

However, this tedious non-linear resolution step can be skipped thanks to Fourier transforms. This technique is summarized in Section 2.6.1. Without solving any equation, one can generate a field $Y(\boldsymbol{x})$ over a unit cell of size $L_1 \times L_2 \times \cdots \times L_d$ where d is the space dimension; d is equal to 2 for fractures. Because of the use of the Fourier transforms, this cell is the unit cell of an infinite spatially periodic medium such as the one displayed in Fig. 2.6. We shall return to this point in Section 2.4.2.

Let us insist again on the importance of linear combinations such as (2.14) since they correspond to a powerful and simple way to generate correlated fields from uncorrelated ones.

2.3.2 Generation of a random fracture

The procedure in the previous subsection can be generalized to generate two random two-dimensional fields $h^+(\boldsymbol{x})$ and $h^-(\boldsymbol{x})$ which correspond to the fluctuations of the upper and lower surfaces of the fracture around their average planes. The components of the two-dimensional vector $\boldsymbol{x}$ are (x, y). First, generate two independent fields $Y_1(\boldsymbol{x})$ and $Y_2(\boldsymbol{x})$. Second, define $h^+ = Y_1(\boldsymbol{x})$ and $h^- = \alpha Y_1(\boldsymbol{x}) + \beta Y_2(\boldsymbol{x})$ where α and β are two coefficients. Then, the two fields $h^+(\boldsymbol{x})$ and $h^-(\boldsymbol{x})$ are correlated as is shown in Exercise 2.3.

2.4 Geometrical properties

This section emphasizes some basic properties of the fractures which were generated in the previous section.

2.4.1 Contact zones

The simulated surfaces $z^+(\boldsymbol{x})$ and $z^-(\boldsymbol{x})$ defined by (2.1) and generated as described in Section 2.3.2 may intersect since the fields $h^{\pm}(\boldsymbol{x})$ are random. This obviously occurs when the roughness σ_h is of the order of $h_o^+ - h_o^-$. This corresponds to regions where

$$z^+(\boldsymbol{x}) \leq z^-(\boldsymbol{x}) \tag{2.20}$$

When this occurs, the local aperture b is equal to 0. Therefore, the definition (2.2) should be made more precise. The aperture b of the fracture is the difference $w = z^+ - z^-$ when it is non-negative

$$b = \begin{cases} w, & w(\boldsymbol{x}) \geq 0 \\ 0, & w(\boldsymbol{x}) < 0 \end{cases} \tag{2.21}$$

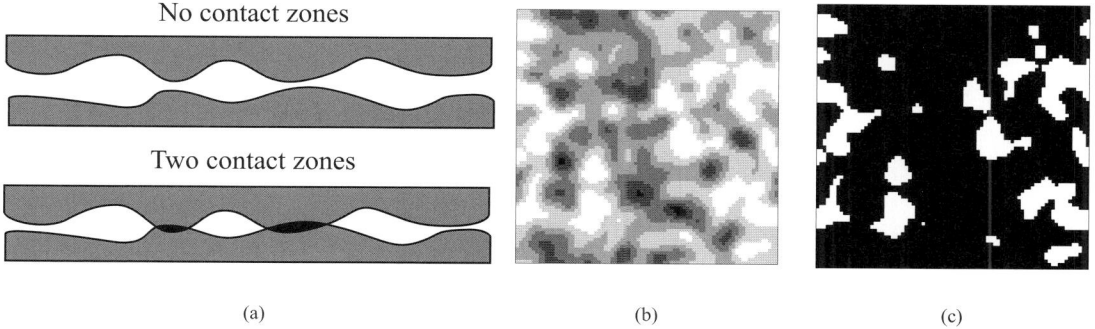

Fig. 2.5 (a) Contact zones; (b) examples of simulated aperture distributions over the fracture plane; (c) in black the corresponding open areas. Six levels of shadings are distinguished in (b), from zero (white) to the largest value (dark) of the aperture. The shading steps are equal to $1.25\sigma_h$. The white areas in (b) and (c) are contact zones. Data are for $b_m/\sigma_h = 1, H = 1$ (b,c). (Reprinted with permission from Mourzenko et al., 1996a).

This is illustrated in several ways in Fig. 2.5.

Of course, real fractures have large portions of contact zones between the two solid blocks located on each side. Let A be the surface of the fracture projected onto the xy-plane. Denote by S_c the proportion of the fracture surface A where the two blocks are in contact. S_c is equal to the average of the phase function Z_c of the contact zones

$$S_c = \frac{1}{A} \int_A Z_c(\boldsymbol{x}) \mathrm{d}s = \overline{Z_c(\boldsymbol{x})} \qquad (2.22\mathrm{a})$$

with

$$Z_c(\boldsymbol{x}) = \begin{cases} 1, & b(\boldsymbol{x}) = 0 \\ 0, & b(\boldsymbol{x}) > 0 \end{cases} \qquad (2.22\mathrm{b})$$

$\mathrm{d}s$ is the differential area element. The fractional open area S_0 where the two blocks are not in contact is trivially

$$S_0 = 1 - S_c \qquad (2.22\mathrm{c})$$

These contact zones make many calculations more complicated. For instance, the average aperture $\bar{b}$ is not equal to b_m as expressed by (2.4); it is precisely

$$\bar{b} = \frac{1}{A} \int_{A_o} b \, \mathrm{d}s \qquad (2.23)$$

where A_o is the open area and A the surface of the fracture projected onto the xy-plane (Fig. 2.5).

Such calculations can be made easily for quantities such as the average fractional void area and the average aperture. The corresponding integrals over the surface are restricted to the areas where $b(\boldsymbol{x}) > 0$ and they were systematically determined by Mourzenko et al. (1999). For instance, for mutually uncorrelated surfaces ($\theta_I = 0$) with Gaussian height distribution (2.5a), the calculations are more easily made by statistical than by spatial averages; therefore,

$$S_0 = 1 - \langle Z_c(\boldsymbol{x}) \rangle = \frac{1}{2}\mathrm{erfc}\left(-\frac{b_m}{2\sigma_h}\right) \quad (2.24\mathrm{a})$$

and

$$\langle b \rangle = \frac{1}{2}\,\mathrm{erfc}\left(-\frac{b_m}{2\sigma_h}\right) b_m + \frac{\sigma_h}{\sqrt{\pi}} \exp\left[-\frac{b_m^2}{4\sigma_h^2}\right] \quad (2.24\mathrm{b})$$

2.4.2 Spatially periodic media

Real fractures may be statistically homogeneous, which means that they are translationally invariant at some scale.

However, fractures which are generated by correlated random fields derived from the Fourier transform technique are generally *spatially periodic*. This means that the medium is supposed to be the juxtaposition of an infinite number of unit cells; these cells can be parallelograms of sides $\boldsymbol{e}$ and $\boldsymbol{f}$ (see Fig. 2.6) with an identical content

$$h^{\pm}(\boldsymbol{x} + \boldsymbol{X}_{n,m}) = h^{\pm}(\boldsymbol{x}) \quad (2.25\mathrm{a})$$

where $\boldsymbol{X}_{n,m}$ is a multiple of the spatial period which can be expressed as

$$\boldsymbol{X}_{n,m} = n\boldsymbol{e} + m\boldsymbol{f} \quad (2.25\mathrm{b})$$

where n and m are a couple of integers. Note that $L_1 = |\boldsymbol{e}|$ and $L_2 = |\boldsymbol{f}|$ (cf. Section 2.3.1). An example of such a spatially periodic medium is shown in Fig. 2.6.

Functions which verify (2.25a), are called spatially periodic. This property is often indicated by the mathematical accent ($\check{}$).

It should be added that in contradiction with the presentation of Section 1.3 permeability is usually calculated on spatially periodic structures such as shown in Fig. 2.6. Permeability is determined as explained in Section 1.3 only if the structure is provided by experiments with no possibility of having spatially periodic boundary conditions. This question is addressed again in Chapter 4.

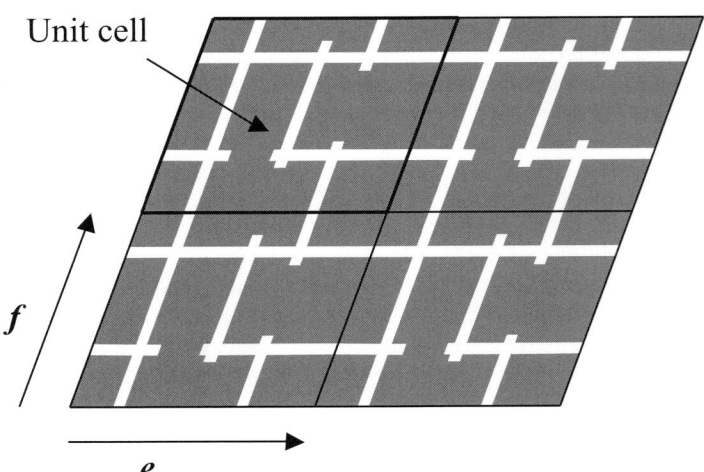

Fig. 2.6 An infinite spatially periodic medium derived by translations from a unit cell.

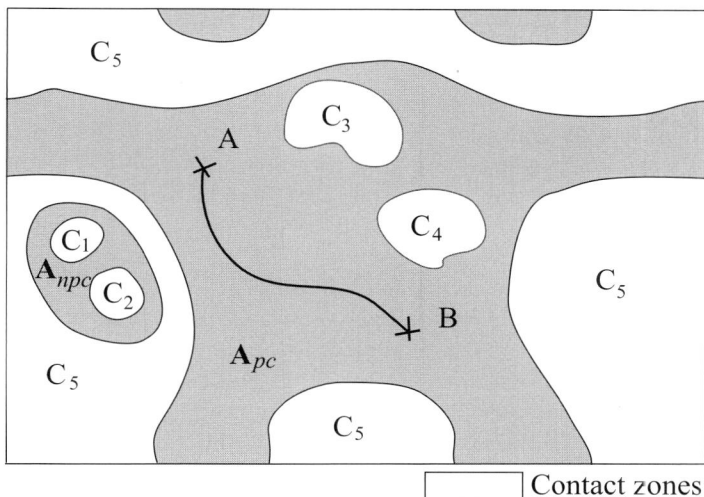

Fig. 2.7 Connected and percolating components in a 2D surface.

2.4.3 Connected and percolating components

Fig. 2.7 displays white and grey areas which can be thought of as closed and open areas of a fracture viewed from above as in Fig. 2.5c.

Two points of a set are *connected* if a continuous path inside the set goes from one point to the other. For instance, in Fig. 2.7, the two points A and B are connected by the solid curve which lies entirely in the subset A_{pc}.

More generally, a *connected component* is the set of points connected to a given point. For instance, suppose that Fig. 2.7 is the unit cell of a spatially periodic fracture. Then, the grey sets comprise two connected components, namely A_{pc} and A_{npc}.

A connected component is a *percolating component* if it goes through the entire sample in at least one spatial direction. For instance, A_{pc} which crosses the whole sample along the horizontal direction, is a percolating component. In contrast, A_{npc} is not a percolating component.

These distinctions are important for flows, for instance. Suppose that a pressure gradient is exerted along the horizontal direction; it will generate some flow in A_{pc}, but not in A_{npc}. Exercise 2.4 illustrates these concepts.

2.5 The concept of percolation

Only a brief outline of percolation is provided in this section. Books have been devoted to this topic such as the one written by Stauffer and Aharony (1994). Sahimi (2011) also provides a short account of this subject. Moreover, this topic has already been discussed at length by Adler and Thovert (1999). The purpose of this brief review is to provide the basic material for an introductory course.

2.5.1 Percolation, cluster and percolation threshold

The starting point of the percolation theory is the classical paper by Broadbent and Hammersley (1957), in which they considered the influence of a random medium on a fluid that flows through it. The medium was modeled as a system of channels that were opened (or closed) with probability P (or $1 - P$). Most early work dealt with the determination of the so-called *critical probability* P_c (or *percolation threshold*); for $P < P_c$, the medium does not percolate; in other words, it is closed. In contrast, for $P \geq P_c$, the medium percolates and a fluid can flow through it when submitted to an external pressure drop.

Let us start with bond percolation. Take a square lattice such as the one displayed in Fig. 2.8a; view it as a conducting mesh where each segment is metallic. These segments are also called bonds. Put the left side at an electrical potential equal to 1 and the right side at a potential 0. An electrical current will flow from left to right. Note that in order to be consistent with the previous paragraph, we could have said that each segment is a circular pipe through which fluid can flow; then, pressure 1 is exerted at the left side of the lattice and pressure 0 at the right side. The two points of view, electrical vs. hydrodynamic, are totally equivalent.

A network is said to percolate if one can go from the left side to the right side by "walking" on the existing bonds. Equivalently, an electrical current or fluid will flow in such a network when it is properly energized.

Now, cut a few bonds (or a few segments) at random with a certain probability $1 - P$. If $1 - P$ is small enough (or equivalently P is still large enough), one obtains a network like that in Fig. 2.8b. It is clear that this network still percolates.

Cut some more bonds as in Fig. 2.8c and current (or fluid) can no longer circulate. The network does not percolate, since there is no continuous path going from one side to the other. The percolation threshold P_c has been crossed.

Though we wish to introduce the minimum number of definitions, we still need the cluster one. *A cluster* is a set of connected bonds as illustrated in Fig. 2.8c.

Of course, one should deal with an infinite network. This infinite network is said to percolate if it contains one infinite cluster, or more loosely if one can go from $-\infty$ to $+\infty$. The probability above which an infinite cluster exists is called the percolation threshold and is denoted by P_c. In most cases, P_c is determined by computer experiments.

The same definitions and concepts can be made on a set of sites with the convention that an occupied site is equivalent to an open bond. Consider a checkerboard and decide at random again whether each square is occupied or not with a probability P as illustrated by Fig. 2.9. An example of percolating cluster is represented in (c).

As mentioned, these percolation probabilities are mostly determined by computer experiments. Some useful results are presented in Table 2.1a.

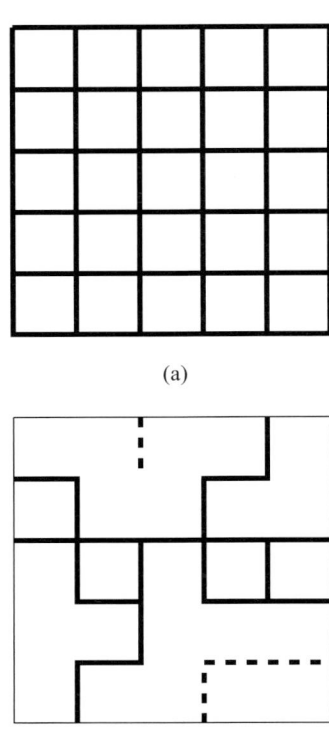
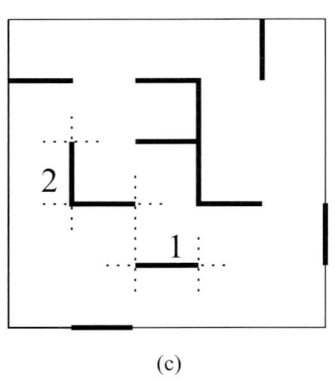

Fig. 2.8 Bond percolation on a square lattice. Open bonds are indicated by thick lines. The thin lines which surround the lattice are added for the sake of clarity; they are not bond. (a) All bonds are opened at $P = 1$. (b) The probability is large enough so that an infinite cluster still exists, together with isolated bonds (broken lines). (c) The probability P is small, and the open bonds form isolated clusters; the empty neighbors of two clusters of sizes 1 and 2 are indicated by portions of dotted lines.

Table 2.1 Percolation thresholds P_c for various lattices (a). Critical exponents for percolation in two and three dimensions (b); rational numbers are assumed exact.

Lattice	Site	Bond
Square	0.593	1/2
Triangular	1/2	0.347
Simple cubic	0.312	0.249
Body-centered cubic	0.246	0.180
Face-centered cubic	0.198	0.119

Exponents	ν	t
d=2	4/3	1.30
d=3	0.88	2.0

(a)　　　　　　　　(b)

(a)

(b)

(c)

Fig. 2.9 Site percolation on a square lattice. The occupation probability is P; occupied sites are black in (a) and (b); in (c), the grey sites are occupied, but do not belong to the percolating cluster represented by the black sites. Data are for: $P = 0.3$ (a), 0.5 (b), 0.62 (c).

2.5.2 Power laws

Let us define the correlation length ξ as the average distance between two sites which belong to the same cluster. ξ can be used to characterize the cluster sizes. Numerical experiments showed that ξ diverges according to a power law close to the percolation threshold

$$\xi \propto |P - P_c|^{-\nu} \text{ for } P \lesssim P_c \qquad (2.26)$$

Therefore, percolation occurs when ξ becomes infinite in agreement with the previous definition. Moreover, one knows how the cluster grows close to percolation. It may be added that above P_c, the previous formula is still valid, but ξ corresponds to the size of the largest finite cluster.

Other macroscopic properties obey power laws as well. Consider, for instance, the conductivity of networks such as the one displayed in Fig. 2.8. Ohm's law is supposed to apply to each bond. Then, the macroscopic conductivity $\langle \Lambda \rangle$ is the statistical average of the conductivity of a sufficiently large number of networks; these computer experiments show that $\langle \Lambda \rangle$ varies as

$$\begin{aligned} \langle \Lambda \rangle &\propto (P - P_c)^t \text{ for } P \gtrsim P_c \\ &= 0 \qquad\qquad \text{for } P < P_c \end{aligned} \qquad (2.27)$$

Since the network does not percolate below P_c, the macroscopic conductivity is obviously equal to zero for $P < P_c$.

A new concept which is specific to the percolation theory (and similar fields) should be introduced here and is called *universality*. The exponents ν and t only depend on the space dimension d and not on the details of the structure. They are the same for both a square and a hexagonal lattice, but they are different for a square and a cubic lattice. Some numerical results are summarized in Table 2.1b.

In contrast to the universal character of the exponents, the percolation thresholds are not universal since they depend on the precise lattice as shown by Table 2.1a; of course, they also depend on d. It is important to note that the proportionality coefficients in (2.26) and (2.27) are not universal as are the percolation thresholds.

2.5.3 Self similarity

An important property of the infinite cluster at percolation is that it is *fractal*. This concept, which was introduced by Mandelbrot in the 1970 can be defined loosely as follows. The cluster is invariant by dilation. This can be understood in simple terms in Fig. 2.10. Consider the whole picture of size L; then isolate, for instance, the highest pyramid of size $L/2$ and dilate it by a factor of 2; you will see precisely the same as in the first picture. And this isolation and dilation process can be continued; take one of the pyramids of size $L/4$ and dilate it by a factor of 4; again one has the same picture. This is exactly the idea behind the fractal character; you have an object, you dilate a part of it and you have the impression, at least in a qualitative statistical sense, that you are looking at the same object. It is due to this fractal character that the macroscopic properties of the percolating structure are power laws. A much more detailed (and historical) introduction is provided by Mandelbrot (1982).

2.5.4 Finite size effects

The existence of the cluster which becomes infinite at P_c implies a crucial feature for both numerical computations and experiments. Let us assume that calculations are done in a cell of size L; in principle, then, L should be very large when compared to any size which is present in the system. But, since ξ tends to infinity, whatever L, then ξ will become comparable to L when P is sufficiently close to P_c. Therefore, the results depend on the ratio L/ξ; this is the so-called *finite size effect*.

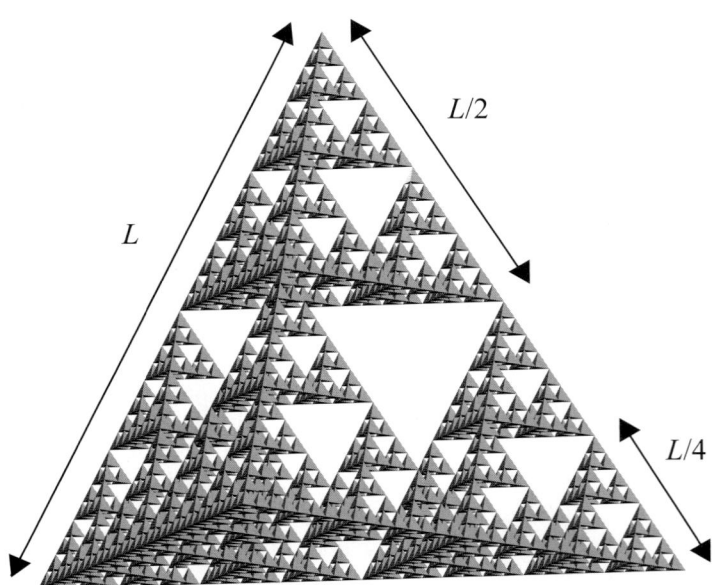

Fig. 2.10 An example of a deterministic fractal structure.

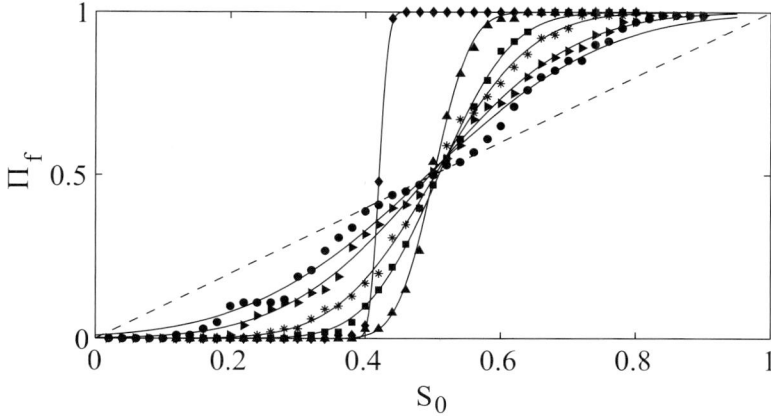

Fig. 2.11 The percolation probability Π_f as a function of the fraction of the open area S_0 for a single fracture discretized by unstructured triangles for various values of ℓ_c/L. Data are for $\ell_c/L = 0$ (♦), 0.0885 (▲), 0.177 (■), 0.354 (∗), 0.71 (▶), 1 (●). The thin solid lines are fits by complementary error functions (cf. eqn 3.7b). The diagonal broken line is the result $\Pi_f = S_0$ when $\ell_c/L \gg 1$.

The finite size of the cell also implies another phenomenon. In principle, an infinite network does not percolate when $P<P_c$, no matter how small $P_c - P$ is. The network percolates with probability 1 when $P \geq P_c$. This is no longer true for a finite network. For example, Fig. 2.11 displays the percolation probability Π_f of a fracture of size $L \times L$ whose Gaussian upper and lower surfaces are correlated by (2.8) which involves the correlation length ℓ_c; such a fracture is illustrated in Fig. 2.5c. Π_f depends on the fraction of open area S_0 and on the ratio ℓ_c/L finite size effects are expected to occur when ℓ_c/L is not sufficiently small. Each curve is obtained by generating a large number of independent networks. Because of this independence, the outcome should follow the law of large numbers and it is usually approximated by a complementary error function erfc (cf. Gradshteyn and Ryzhik, 1965). Because of the finite size effect, Π_f is smoothed out when ℓ_c/L is not sufficiently small. But the standard deviation goes to zero when ℓ_c/L decreases and Π_f becomes progressively closer and closer to a step function. We shall return to these points in Section 3.4.3.

One simple way to obtain the properties for infinite cells is to proceed by extrapolation. Calculations are made for several cell sizes L and one obtains, for instance

$$<\Lambda> = f(P, L) \quad (2.28a)$$

This is usually extrapolated to infinity by plotting $<\Lambda>$ as a function of L^{-1} in some adequate coordinates. The value for $L = \infty$ is obtained by the extrapolated ordinate at the origin.

For instance, at the percolation threshold, according to Fischer (1971), $<\Lambda>$ is expected to verify

$$<\Lambda> \propto L^{-t/\nu} \quad (2.28b)$$

2.5.5 The percolation thresholds for fractures

As in three dimensions for porous media, the correlated character or not of the fractures plays an important role for the percolation threshold.

Systematic estimations of the threshold for fracture surfaces uncorrelated and correlated by the Gaussian correlation function (2.8) were done by Mourzenko et al. (1996a).

For uncorrelated surfaces, they obtained a threshold which corresponds to a fractional open area S_0 equal to 0.59 in agreement with the results contained in Table 2.1. For correlated fractures, the threshold is equal to 0.5 in agreement with the literature results too.

2.6 Extensions

This section briefly addresses the practical generation of correlated fields in Section 2.6.1; these formulae should be complete and ready to use. Then, Section 2.6.2 provides a general presentation of self-affine surfaces; they are defined and their power spectral density is shown to be a power law whose exponent is related to the Hurst exponent (cf. eqn 2.42); they can be generated by the technique described in Section 2.6.1; finally, some of their most important statistical properties are provided without proof.

2.6.1 Generation of correlated fields by Fourier transforms

Adler (1992) gave a full account of the technique of Fourier transforms. Adler and Thovert (1999) summarized it as follows for a square unit cell of size L. This technique is very powerful and avoids the resolution of the non-linear equations (2.19). For instance, the field is supposed to be correlated according to (2.8) or (2.10) with a correlation length ℓ_c.

Consider a unit cell composed of $N_L \times N_L$ elementary squares of size $\Delta x = \Delta y = a$; the correlation length ℓ_c is discretized into n_ℓ such squares, i.e. $\ell_c = n_\ell a$, $L = N_L a$. At each node $(x_m = ma, y_n = na)$, the Gaussian spatially correlated periodic field Y_{mn}, which will stand for h^+ or h^-, can be calculated as

$$Y_{mn} = N_L \sum_{s,t=0}^{N_L-1} \sqrt{\hat{C}_{st}} \, \hat{X}_{st} \exp\left[-\frac{2\pi i(ms+nt)}{N_L}\right], \quad m,n = 0,\ldots,N_L-1 \tag{2.29}$$

where $\hat{X}_{st}$ is the Fourier transform of a non-correlated standard Gaussian field X_{mn}

$$\hat{X}_{st} = \frac{1}{N_L^2} \sum_{m,n=0}^{N_L-1} X_{mn} \exp\left[\frac{2\pi i(ms+nt)}{N_L}\right] \tag{2.30a}$$

$\hat{C}_{st}$ is the spectral density function which is by definition the Fourier transform of the autocorrelation function; it is generally denoted by $\mathcal{S}_{st}$

$$\mathcal{S}_{st} = \hat{C}_{st} = \frac{1}{N_L^2} \sum_{m,n=0}^{N_L-1} C_{mn} \exp\left[\frac{2\pi i(ms+nt)}{N_L}\right], \quad s,t = 0,\ldots,N_L-1 \tag{2.30b}$$

C_{mn} is the covariance matrix of Y_{mn} which is derived from (2.8) or (2.10) for instance

$$C_{mn} = C_h\left(\sqrt{x_m^2 + y_n^2}\right) \quad (2.31)$$

Note that because of (2.31) the field Y has been implicitly multiplied by a factor σ_h. This implies that Y is Gaussian, with a zero mean and a variance equal to σ_h^2.

As stated, this Fourier transform technique is equivalent to a linear combination with the crucial advantage that the coefficients of the linear combination are easily deduced from the autocorrelation function. It is easily extended to the anisotropic correlation functions (2.9) and (2.11).

An important property of the spectral density is given by the Wiener–Khintchine theorem which can be stated as follows for products of spatially periodic functions $\check{Y}$ and $\check{Z}$

$$C_{\check{Y}\check{Z}}(\boldsymbol{r}) = \frac{1}{\tau_0}\int_{\tau_0} \check{Y}(\boldsymbol{r}_1)\check{Z}(\boldsymbol{r}+\boldsymbol{r}_1)\mathrm{d}^d\boldsymbol{r}_1 \quad (2.32\mathrm{a})$$

where d is the dimension of the domain τ_0. Then,

$$\hat{C}_{\check{Y}\check{Z}}(\boldsymbol{k}) = \hat{\check{Y}}_{\boldsymbol{k}}\,\hat{\check{Z}}^*_{\boldsymbol{k}} \quad (2.32\mathrm{b})$$

where the star denotes the complex conjugate and $\boldsymbol{k}$ the reciprocal vector.

2.6.2 Self-affine surfaces

Definition

Self-affine surfaces can be considered as the opposite of the Gaussian correlated surfaces (cf. 2.8) since they are statistically invariant by some dilation while the second ones are invariant by translation (cf. Subsection 2.5.3).

A function $z(x)$ is exactly *self-affine* (Mandelbrot, 1982) if it is invariant by the transformation

$$z(\alpha x) = \alpha^H z(x) \quad (2.33)$$

where α is an arbitrary real number.

Of course, for applications to real objects, this exact definition should be extended in a statistical sense. More precisely, the random curve $Z(x)$ is a self-affine fractal if

$$\mathrm{Prob}\{\frac{Z(x+u)-Z(x)}{u^H} \leq \Delta\} = F(\Delta) \quad (2.34)$$

where $F(\Delta)$ can be any probability distribution function, though most of the time, $F(\Delta)$ is the Gaussian distribution function. $\mathrm{Prob}(\omega)$ denotes the probability of the event ω.

Spectral properties

There are several ways to derive the spectral properties of self-affine curves. A qualitative demonstration based on (2.33) and due to Turcotte (1992) is proposed. Consider a curve $z_1(x)$ where x varies between 0 and L. The corresponding finite Fourier transform is expressed as

$$\hat{z}_1(k, L) = \int_0^L z_1(x) e^{2i\pi k x} \, dx \qquad (2.35)$$

According to the Wiener–Khintchine theorem (2.32b), the power spectral density defined as the Fourier transform of the autocorrelation function is

$$\mathcal{S}_1(k) = \frac{1}{L} \hat{z}_1(k, L) \, \hat{z}_1^*(k, L) \qquad (2.36)$$

Consider a second curve $z_2(x)$ which is related to $z_1(x)$ by the self-affine relation (2.33)

$$z_2(x) = \frac{1}{r^H} z_1(rx) \qquad (2.37)$$

An elementary change of variable of x into rx shows that

$$\hat{z}_2(k, L) = \frac{1}{r^{H+1}} \hat{z}_1\left(\frac{k}{r}, rL\right) \qquad (2.38)$$

Thus, the spectral density $\mathcal{S}_2(k)$ of $z_2(x)$ is easily deduced as

$$\mathcal{S}_2(k) = \frac{1}{r^{2H+1}} \frac{1}{rL} \left|\hat{z}_1\left(\frac{k}{r}, rL\right)\right|^2 = \frac{1}{r^{2H+1}} \mathcal{S}_1\left(\frac{k}{r}\right) \qquad (2.39)$$

Suppose now that k and r are chosen so that

$$k = rk_0, \qquad k_0 = \text{const} \qquad (2.40)$$

(2.39) can be modified as

$$\mathcal{S}_2(k) = \frac{k_0^{2H+1}}{k^{2H+1}} \mathcal{S}_1(k_0) \qquad (2.41)$$

Hence, the spectral density $\mathcal{S}_2(k)$ behaves like a power law

$$\mathcal{S}_2(k) \propto k^{-(2H+1)} \qquad (2.42)$$

This physical argumentation has the great advantage of yielding the correct result (2.42) with a minimum effort. There are other more precise ways to obtain this result, which are cited by Adler and Thovert (1999) who also provide additional properties of self-affine fields.

Let us go back for a minute to the simple extension of the Gaussian correlation which is expressed by (2.10) and which was introduced for pedagogical purposes. A Taylor series expansion of this relation in terms of the ratio u/ℓ_c shows that the leading term corresponds exactly to a self-affine surface with Hurst exponent H. Hence, (2.10) is a good approximation to a self-affine surface in the range $u < \ell_c$.

Generation

Let us now turn to the practical generation of self-affine surfaces and to some of their properties. The previous developments can be summarized by the following two formulae. Their covariance C_h can be characterized by their Fourier spectrum

$$C_h(\mathbf{u}) = \int \mathcal{S}(\mathbf{k}) e^{-2i\pi \mathbf{k}\cdot \mathbf{u}} \mathrm{d}^2 \mathbf{k}, \quad \mathcal{S}(\mathbf{k}) \sim k^{-2H-2}, \quad 0<H<1 \quad (2.43\mathrm{a})$$

The random fields h^+ and h^- are generated by the standard method of Fourier transforms by imposing their variance σ_h^2 and the power spectrum of the covariance C_h

$$\frac{\mathcal{S}(\mathbf{k})}{\mathcal{S}(0)} = \begin{cases} \|\mathbf{k}\Lambda_{co}\|^{-2H-2}, & \|\mathbf{k}\Lambda_{co}\| > 1 \\ 1, & \|\mathbf{k}\Lambda_{co}\| \leq 1 \end{cases} \quad (2.43\mathrm{b})$$

The normalization constant $\mathcal{S}(0)$ is set so that the integral of the power spectrum equals the variance σ_h^2. The length scale $\Lambda_{co} = n_\Lambda a$ is the upper cut-off for the self-affinity of the surface profiles. Note that the exponents of k in (2.42) and (2.43) differ by 1 because these formulae apply to lines and surfaces, respectively.

Practically, the fractures are generated as follows. Square fractures of size $L \times L$ are reconstructed by generating the heights h^+ and h^- at the nodes $(i\Delta x, j\Delta y)$ of a regular square grid with $\Delta x = \Delta y = a$, $L = N_c a$ and $i, j = 1, 2, ..., N_c$. These fractures are spatially periodic, and Λ_{co} should verify that $\Lambda_{co} \leq L/2$.

A square domain of size $\lambda = n_\lambda a$ is then cut from this master sample. The separation b_m is adjusted so that the fractional oper. area in the subdomain is equal to some prescribed value. Self-affine behavior is expected only for a range of domain sizes large enough to encompass several characteristic scales, but still much smaller that the cutoff length Λ_{co}

$$1 \ll n_\lambda \ll n_\Lambda \quad \text{or} \quad a \ll \lambda \ll \Lambda_{co} \quad (2.44)$$

Statistical properties

For certain random fields, volume averages are equivalent to statistical averages. This property, known as ergodicity, holds for most fields in this book. However, it does not hold for self-affine fields and the two possible averages should not be confused.

For self-affine fractures, many average properties depend on the size of the domain in contrast with statistically homogeneous fractures. Therefore, subdomains Ω of a fracture F with size λ are introduced. The spatial averages and variances over such domains are denoted by an overbar $\overline{(.)}$ in accordance with the notations given in Section 2.2.1

$$\overline{X} = \frac{1}{\Omega} \int_\Omega X(\mathbf{x}) \mathrm{d}^2 \mathbf{x} \quad \overline{\sigma_X}^2 = \frac{1}{\Omega} \int_\Omega \left(X(\mathbf{x}) - \overline{X}\right)^2 \mathrm{d}^2 \mathbf{x} \quad (2.45)$$

For instance, the average and variance of the aperture over Ω are $\bar{b}$ and $\overline{\sigma_b}^2$, respectively.

Conditional averages of $\overline{X}$ over domains which share a common value y of some random variable Y are denoted by $\langle \overline{X} \rangle_Y$, i.e.

$$\langle \overline{X} \rangle_Y (y) = \frac{1}{N_y} \sum_{i=1}^{N_y} \overline{X}_i \qquad (2.46)$$

where N_y is the number of domains where the variable Y is equal to y. For instance, $\langle \bar{b} \rangle_{\overline{w}/\overline{\sigma_h}}$ is the mean local aperture of all fracture samples which have a given ratio $\overline{w}/\overline{\sigma_h}$. Recall that w is defined in Section 2.4.1.

Mourzenko et al. (1999) derived many geometrical properties of self-affine fractures and some of them are recalled here without any proof. For any finite domain with size $\lambda \ll \Lambda_{co}$, cut from a larger fracture generated as described before, the mean separation $\overline{w}$, the mean aperture $\bar{b}$ and the variances $\overline{\sigma_b}^2$ and $\overline{\sigma_h}^2$ of the aperture and surface elevation are defined from simple spatial statistics, denoted by the overbars. The same is true for the fractional open area $S_{0,\lambda} = 1 - \overline{Z_c(\boldsymbol{x})}$ measured over a domain of size λ.

The statistical averages of $\overline{w}$, $\bar{b}$ and $S_{0,\lambda}$ do not depend on the sample size λ, and they are obviously equal to the spatial averages over large domains with respect to the cut-off length Λ_{co}

$$\langle \overline{w} \rangle = b_m \qquad \langle \bar{b} \rangle = \langle b \rangle \qquad \langle S_{0,\lambda} \rangle = S_0 \qquad (2.47)$$

However, due to the self-affine character of the fracture, the variances $\overline{\sigma_b}^2$ and $\overline{\sigma_h}^2$ are expected to increase with λ. Mourzenko et al. (1999) showed that their expectations Σ_b^2 and Σ_h^2 scale according to the power laws

$$\Sigma_h^2 = \langle \overline{\sigma_h}^2 \rangle \approx \sigma_h^2 \, \Theta(H) \, Q(H) \left(\frac{\lambda}{\Lambda_{co}} \right)^{2H_h}, \quad \left(\frac{\lambda}{\Lambda_{co}} \right)^{2-2H} \ll 1 \quad (2.48a)$$

$$\Sigma_b^2 = \langle \overline{\sigma_b}^2 \rangle \approx \qquad 2 S_0 \Sigma_h^2 \qquad , \quad \left(\frac{\lambda}{\Lambda_{co}} \right)^{2H} \ll 1 \quad (2.48b)$$

where the functions Θ and Q depend on H only. The exponent H_h is close to H, but shifted slightly towards $1/2$ (H_h=0.30, 0.50 and 0.79, for H=0.25, 0.50 and 0.87, respectively), due to the band-limited character of the spectral density function (2.43b) as shown by Kant (1996) and by Mourzenko et al. (1999).

As a consequence of the Gaussian character of the surface heights, the statistical properties of finite pieces of fractures with a prescribed ratio of local mean separation and surface roughness $\overline{w}/\overline{\sigma_h}$ are expressed as

$$\langle S_0 \rangle_{\frac{\overline{w}}{\overline{\sigma_h}}} = \frac{1}{2} \operatorname{erfc} \left(-\frac{\overline{w}}{2\overline{\sigma_h}} \right) \qquad (2.49a)$$

$$\langle \frac{\bar{b}}{\overline{\sigma_h}} \rangle_{\frac{\overline{w}}{\overline{\sigma_h}}} = \langle S_0 \rangle_{\frac{\overline{w}}{\overline{\sigma_h}}} \frac{\overline{w}}{\overline{\sigma_h}} + \frac{1}{\sqrt{\pi}} \, e^{-\frac{\overline{w}^2}{4\overline{\sigma_h}^2}} \qquad (2.49b)$$

$$\langle \frac{\overline{\sigma_b}^2}{\overline{\sigma_h}^2} \rangle_{\frac{\overline{w}}{\overline{\sigma_h}}} = \langle S_0 \rangle_{\frac{\overline{w}}{\overline{\sigma_h}}} \left(2 + \frac{\overline{w}^2}{\overline{\sigma_h}^2} \right) + \frac{\overline{w}}{\sqrt{\pi} \, \overline{\sigma_h}} \, e^{-\frac{\overline{w}^2}{4\overline{\sigma_h}^2}} - \langle \frac{\bar{b}}{\overline{\sigma_h}} \rangle_{\frac{\overline{w}}{\overline{\sigma_h}}}^2 \qquad (2.49c)$$

Furthermore, the domain size λ has absolutely no influence; whichever λ/Λ_{co}, $\langle \overline{b}/\overline{\sigma_h} \rangle_{\overline{w}/\overline{\sigma_h}}$ or $\langle \overline{\sigma_b}^2/\overline{\sigma_h}^2 \rangle_{\overline{w}/\overline{\sigma_h}}$ are identical for identical $(\overline{w}/\overline{\sigma_h})$.

Therefore, the statistical geometry of subdomains of a fracture is fully characterized by the two parameters $\overline{\sigma_h}$ and $\overline{w}/\overline{\sigma_h}$, with all the scale effects embodied in the scaling law (2.48a) for $\overline{\sigma_h}$. Alternatively, $\overline{\sigma_b}$ and $\overline{b}/\overline{\sigma_b}$ or S_0 can be used instead of $\overline{\sigma_h}$ and $\overline{w}/\overline{\sigma_h}$ to describe the geometry, since the two sets of parameters are related by (2.49).

Finally, Mourzenko et al. (1999) considered the percolation probability $\Pi_f(S_{0,\lambda}, \lambda)$ of finite samples of self-affine fractures of size λ and of open surface $S_{0,\lambda}$. The most striking observation is that $\Pi_f(S_{0,\lambda}, \lambda)$ is actually independent of the sample size λ, provided it is in the self-affine range, $a \ll \lambda \ll \Lambda_{co}$. For H=0.25, 0.50 and 0.87, the critical fractional open area is equal to 0.57, 0.53 and 0.535, respectively. Therefore, the influence of H is relatively small.

Exercises

(2.1) Evaluate I_h for uncorrelated surfaces and for perfectly correlated surfaces.

(2.2) Demonstrate the first equality in (2.15).

(2.3) Let $Y_1(\boldsymbol{x})$ and $Y_2(\boldsymbol{x})$ be two independent fields which verify $\langle Y_i \rangle$=0 and $\langle Y_i^2 \rangle = \sigma_h^2$ for i=1,2.

 (i) Calculate the average and the variance of the field $h^- = \alpha Y_1(\boldsymbol{x}) + \beta Y_2(\boldsymbol{x})$ with $\alpha^2 + \beta^2 = 1$.
 (ii) Calculate the intercorrelation coefficient θ_I of the two surfaces $h^+ = Y_1(\boldsymbol{x})$ and h^-.
 (iii) What are the consequences?
 (iv) Of course, so far we have only been interested in the fluctuations around the average planes. The two real surfaces are going to be expressed by $z^\pm(\boldsymbol{x}) = h^\pm(\boldsymbol{x}) + h_o^\pm$ with a local aperture $b = z^+(\boldsymbol{x}) - z^-(\boldsymbol{x})$. Calculate the intercorrelation between the real surfaces $z^+(\boldsymbol{x})$ and $z^-(\boldsymbol{x})$.

(2.4) Consider the white and grey zones in Fig. 2.7.

 (i) Suppose that the medium is not spatially periodic. Determine the connected components and the percolating components of the grey zones.
 (ii) Determine the connected components and the percolating components of the white zones when the medium is spatially periodic or not.

3 The geometry of fracture networks

3.1 Introduction ... 30
3.2 Classical analysis of a fracture network ... 32
3.3 Generation of a fracture network ... 35
3.4 Statistical geometrical properties of fracture networks ... 40
3.5 Estimation of the dimensionless density from line data ... 48
3.6 Estimation of the dimensionless density from surface data ... 50
3.7 Stereological relations for convex fractures ... 51
3.8 Extensions ... 54
Exercises ... 65

3.1 Introduction

In the previous chapter, fractures were considered at a scale of the order of magnitude of the aperture, i.e. of the order of b_m; therefore, the two solid surfaces limiting the fractures could be distinguished as in Fig. 2.2. In the present chapter, fractures are considered at a scale of the order of their lateral extension which is often denoted by $2R$ (cf. Fig. 3.1a). According to Section 1.2, b_m is of the order of 0.1 mm when R is of the order of a few meters. Therefore, at a scale of the order of R, a fracture looks like a single surface, since one cannot distinguish anymore between the two surfaces which limit the void space.

This single surface may have a plane polygonal shape such as a hexagon. Generally speaking, the fracture is not necessarily plane; however, the radius of curvature is probably intermediate between b_m and R. These two cases are illustrated in Fig. 3.1. Note that the fractures which are not plane, are often either veins which developed in a medium undergoing ductile deformation or the junction of two fractures with slightly different orientations. It will be seen in Chapter 5 that the fractures need not to be plane; as soon as they can be triangulated, the properties of the networks can be calculated.

One of the basic quantities that one would like to know is the fracture density, ρ, i.e. the number of fractures per unit volume. However, the rocks have a trivial property, namely they are not transparent! Therefore, one cannot see how numerous the fractures are, how they are organized and how they cut the rocks into blocks. Because of this obvious fact, fractures are most often seen when they intersect a surface such as an outcrop or a quarry; these intersections are called *traces*. Examples, at different scales, are given in Fig. 3.2.

On the meter scale, a block of dark grey Hercynian granite (about $52 \times 35 \times 36$ cm^3) was analyzed in detail by Ledésert *et al.* (1993). Trace maps were drawn from nine sections of the block and one of them is displayed in Fig. 3.2a. The fracture pattern appeared to be composed of two main families A and B, at about $\pm 30°$ inclination angle from the vertical axis. A three-dimensional representation of the whole network was shown in Fig. 1.3. It is also seen that the traces are not perfectly straight, indicating that the corresponding fractures are not plane.

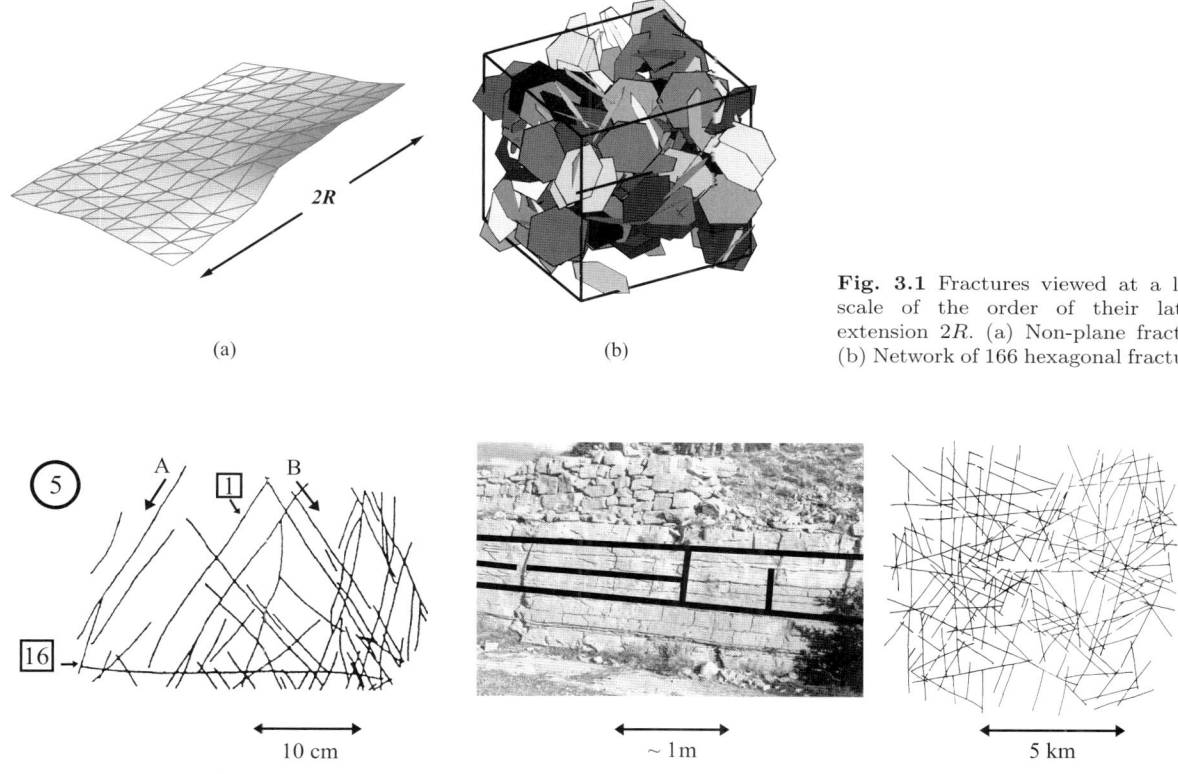

Fig. 3.1 Fractures viewed at a large scale of the order of their lateral extension $2R$. (a) Non-plane fracture. (b) Network of 166 hexagonal fractures.

Fig. 3.2 Traces on plane surfaces. (a) Trace map in one of the sections of the granite block of size $(50 \text{ cm})^3$ measured by Ledésert *et al.* (1993). (b) The main traces of Fig. 1.1b are indicated by the thick black lines. (c) Aerial view of a region in Central Africa reported by Vignes-Adler *et al.* (1991).

On the decameter scale, Fig. 3.2b displays some of the traces already seen in Fig. 1.1b, but now marked by thick lines.

On the large scale ≈ 1 km, Fig. 3.2c is an aerial view of a region in Central Africa. Note that the traces may be masked if, for instance, the ground surface is covered by soil or a thick weathered layer, but the absence of visible traces does not necessarily mean that there are no fractures.

These three maps show that fractures are present at all scales underground. Smaller and larger fractures could be added to the previous list.

This introduction should mention traces recorded along observation lines. An approximate example of such measurements consists of well logs where the intersections of the fractures with the well can be systematically recorded and analyzed; therefore, the well serves as the observation line in this example. Otherwise, observation lines can be drawn on outcrops and the intersections of the traces with such lines can be analyzed as well.

This chapter is organized as follows. The classical geometrical measurements performed on fracture networks are described in Section 3.2. Some of the possible ways to numerically generate fracture networks from data are summarized in Section 3.3. Important statistical properties of fracture networks, such as the percolation threshold, are presented in Section 3.4, together with the concept of excluded volume and of the dimensionless fracture density ρ'. Then, ways to estimate ρ' from line or surface data are detailed in Sections 3.5 and 3.6. Stereological relations are surveyed in Section 3.7. Finally, important extensions to complex networks are summarized in Section 3.8.

3.2 Classical analysis of a fracture network

3.2.1 Analysis of traces on plane surfaces

Suppose that one has an outcrop (or a cliff or quarry) which is approximately plane and that traces are visible on this surface.

Various measurements can be made on this set of traces which are basically lines (more or less straight) on a plane Π as displayed in Fig. 3.3. First, the number of intersections of the traces with a line parallel to the unit vector $\boldsymbol{p}$ drawn on the outcrop can be determined; this number is equal to six in the figure; let $n_I(\boldsymbol{p})$ be the *number of such intersections per unit length of the line*. Then, one can count the total number of traces which are observed and divide this number by the surface of the outcrop to obtain the *surface density of traces* Σ_t, or equivalently the number of traces per unit surface (this number is nine in Fig. 3.3). Another quantity is the *number of intersections between traces per unit surface* denoted by Σ_p; there are nine such intersections in Fig. 3.3.

One can also measure the orientation of the traces with respect to a given direction (north for instance, if the plane surface is horizontal). Sometimes, the inclination of the fracture plane relative to the surface of the outcrop is also recorded. Notations are illustrated in Fig. 3.4a. The plane (x, y) may be thought to be a horizontal plane with the y-axis pointing north. The *strike* is the trace corresponding to the intersection of the fracture with a horizontal plane; let $\boldsymbol{t}$ and $\boldsymbol{p}_f$ be the unit vectors parallel and perpendicular to this line; the azimuth ϕ_a is the

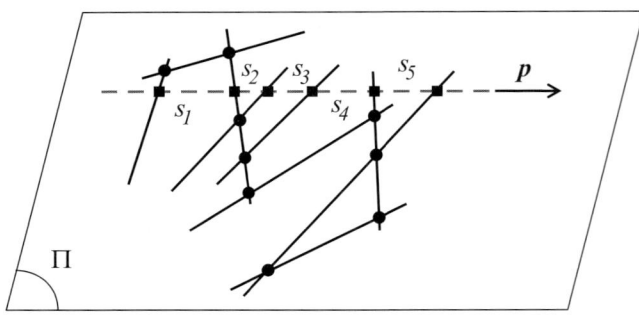

Fig. 3.3 Traces on an outcrop schematized by plane Π. The thick black lines are traces. The bullets (●) are the intersections between the fractures. A broken observation line parallel to the unit vector $\boldsymbol{p}$ is drawn on the outcrop; the black squares (■) are the intersections of the fractures with the line. The spacings s_i ($i = 1, \ldots, 5$) between the intersections of the traces with the observation line are indicated.

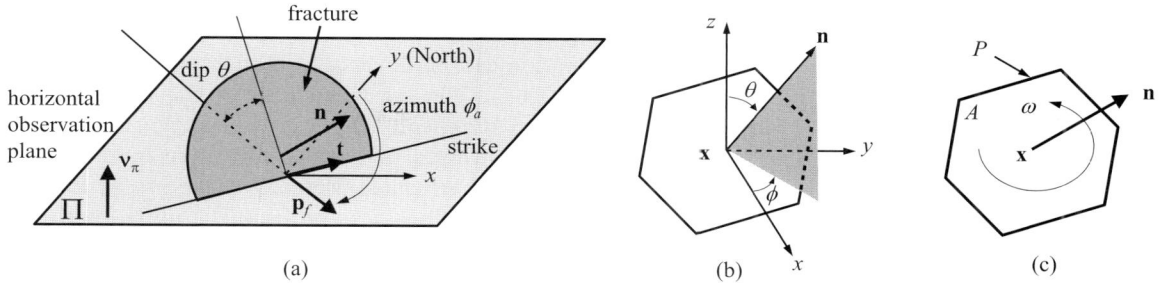

Fig. 3.4 Orientation of a fracture in space. (a) Schematic intersection of a circular fracture and an horizontal plane Π. (b) The polar coordinates (θ, ϕ) of a unit vector $\boldsymbol{n}$ normal to a fracture are displayed in (a); the z-axis is vertical; the y-axis points north. In (c), the orientation of a polygon in space is characterized by the unit normal $\boldsymbol{n}$ and the angle ω.

angle counted clockwise from the north, i.e. the y-axis, to $\boldsymbol{p}_f$; the strike azimuth is the direction of the strike counted clockwise from the north; in our case, it is equal to $\phi_a - \pi/2$. The *dip* is the magnitude of the angle between the fracture and the horizontal plane and is thus equal to θ.

There are several conventions for θ and ϕ_a and we have chosen not to detail them. In our view, it is better to convert them and to calculate the components of the unit normal $\boldsymbol{n}$ to the fracture, or to use the standard polar angles as shown in Fig. 3.4b.

The other standard situation is a vertical outcrop where the y-axis can be chosen as the vertical axis. The same remarks apply.

A final remark is essential. In most cases, i.e. when the fracture is not circular, $\boldsymbol{n}$ is not sufficient to define its orientation in space. One needs another angle, say ω, to specify the polygon orientation within its plane, relative to any arbitrary origin, as illustrated in Fig. 3.4c; ω is called the *in-plane orientation*.

One can also build a histogram of the trace lengths. Let $g(c)$ be the probability density of these lengths; here, c stands for *chords*, a word equivalent to traces. Therefore, $g(c)\mathrm{d}c$ is the probability of finding a trace with a length of between c and $c + \mathrm{d}c$.

Finally, one may also be interested in the distribution of the fractures in space, irrespective of their orientations. A standard way to characterize this distribution is to measure the distribution of the spacings s_j ($j = 1, \ldots, 5$) along an observation line, as illustrated in Fig. 3.3. The spacing histogram and the correlation between two successive spacings are common ways to determine if the traces are statistically independent, in other words, whether the fractures are distributed according to a homogeneous Poisson process or not.

The previous characteristics are only provided as guidelines since each fracture field is unique. Therefore, in each field, one will probably make specific measurements which could be of less interest in another field. As a rule, flexibility and adaptability are essential features of the experimentalist who undertakes measurements in a fracture field (and for the modelist who analyzes the data as well!).

3.2.2 An example of three-dimensional field data

The previous subsection can be illustrated by field measurements made by Sisavath et al. (2004). The karstified part of the Baget watershed located inside the north Pyrenean zone was studied.

Open crack orientations were mapped along three roads; these line surveys correspond to sub horizontal profiles, referred to as P1, P2 and P3, oriented 110°, 70° and 120° clockwise from the north, with lengths 700 m, 200 m and 100 m, respectively. In addition, fractured zones consisting of series of subparallel cracks were characterized by their width and their mean crack spacing.

The orientational distribution of the events is shown in Fig. 3.5a. The orientations were measured with a 5° resolution, but they were slightly randomized in Fig. 3.5a in order to distinguish events with identical orientations. Only very limited information regarding the event extensions can be gained from the present line surveys, since the observed trace lengths are very often truncated by the boundaries of the outcrop along the roads. Fig. 3.5a clearly shows that the events can be categorized into four main families, which are referred to hereafter as F1 to F4. Two of them (F1 and F4) are sub vertical, and roughly in the E-W and N-S directions, whereas F2 and F3 have a slope of about 40°–50°. In addition, a few events do not belong to any of the four families.

Although the data set is too limited for a detailed analysis, some information can be obtained from the 13 events of family F1 on profile P1. The spacings s_j between successive events j and $j+1$ have an average $<s> = 43.3$ m and a standard deviation $\sigma_s = 48.6$ m. When the events are Poissonian, i.e. without any spatial correlation, the spacing between their intersections with a scan line obeys an exponential probability law which is proportional to $\exp(-s/<s>)$, with $<s> = \sigma_s$. In the present case, σ_s is slightly larger than the mean spacing, which indicates that the events are slightly more clustered than in a random distribution. The

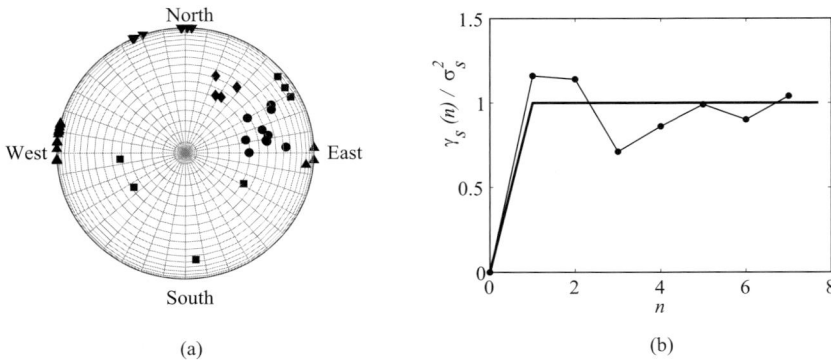

Fig. 3.5 Fracture measurements in the Baget watershed. (a) Orientation of the fracturation events. Symbols correspond to the position of the vector n on the unit sphere. Families F1, F2, F3, and F4 are denoted by ▲, ●, ♦ and ▼, respectively. The line spacing is 10° for the radial angle and 5° for the inclination. (b) Variogram (3.1) of the spacings of the intersections of events from family F1 with profile P1(–●–). The thick solid line corresponds to uncorrelated event locations.

semivariogram of the spacings can be defined as the average as proposed by Yaglom (1957)

$$\gamma_s(n) = \frac{1}{2}\langle (s_j - s_{j+n})^2 \rangle \qquad (3.1)$$

For a Poisson distribution, the spacings are uncorrelated and γ_s is constant and equal to σ_s^2. The variogram γ_s is plotted in Fig. 3.5b for family F1 in profile P1. Deviations from σ_s^2 are observed which are due in most part to the small size of the statistical data set.

No intercorrelation was found, either between the event width and location, or the spacing.

3.3 Generation of a fracture network

There are many different techniques to numerically generate a fracture network and the following only gives a short overview of the various possibilities. Let us again emphasize the necessity for the modelist to listen carefully to the geologist and to generate a network whose characteristics are as close as possible to the specifications.

3.3.1 Complete analysis

This subsection shows how one can fully measure a natural fracture block and the amount of work which this necessitates. To the best of our knowledge, no other comparable study has been reported in the literature.

A block of dark grey Hercynian granite was extracted at La Peyratte, Deux-Sèvres, France. It is fine-grained (1 to 2 mm long crystals) and is crosscut by numerous fractures surrounded by discolored alteration halos. The primary acquisition was undertaken by Ledésert et al. (1993). The granite block (about $52 \times 35 \times 36$ cm^3) was sawed into nine parallel plates, 4 cm in thickness. Trace maps were drawn from the nine sections as illustrated in Fig. 3.6. The fracture pattern appeared to be composed of two main families A and B, at about $\pm 30°$ inclination angle from the vertical axis in Fig. 3.6. In addition to these two families, there is one horizontal fracture (fracture number 16, in Fig. 3.6).

The fracture traces in each section were labeled, and the traces of a given fracture in successive sections were given identical labels. The rock sample contained about 240 fractures. About 150 of them could be seen in a single plane. Their trace lengths ranged between 1 and 12 cm. The other 90 fractures could be followed through at least two planes, with trace lengths between 2.5 and 47 cm. Only six fractures could be traced through the nine sections. The total surface area of fractures was estimated to be 2.6 m^2, and the block volume 0.066 m^3.

The three-dimensional reconstruction was performed by triangulating the fracture surfaces (Gonzalez-Garcia et al., 2000). As shown in Fig. 3.6, the fracture traces in each section are often curved, and the fractures are also often twisted between successive sections.

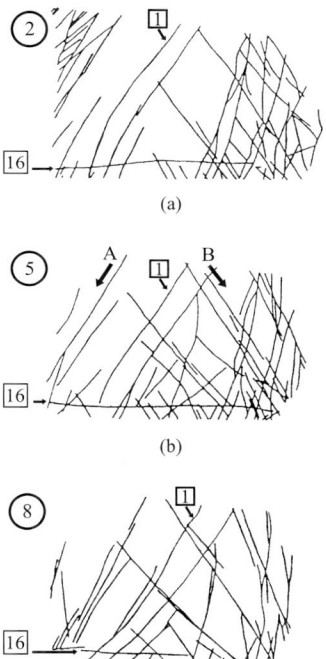

Fig. 3.6 Three of the nine successive trace maps. The directions of the two families A and B are given. The traces of two fractures (1 and 16) are indicated in each section. (Reprinted with permission from Ledésert et al. (1993).)

The triangulation procedure comprises three main steps. In the first step, mesh points are distributed along each fracture trace. This operation is partly manual, but is computer-aided. In the second step, each fracture is separately triangulated, by joining the points which belong to its successive traces. These two steps are conceptually simple, and they are also sufficient to allow three-dimensional visualizations of the network. The third step in the triangulation is by far the most difficult, namely the derivation of the fracture intersections.

Finally, the triangulation of the fracture network is provided by a list of about 3000 mesh points defined by their coordinates, and a list of about 9000 triangles defined by their three vertices. The triangulated network was displayed in Fig. 1.3. Fractures which do not intersect the side boundaries of the rock sample cannot be detected without actually dismantling the block, even though some of them are quite extensive.

Such a set of data means that many studies are possible. However, it is clear that the extraction of all the information contained in a single and relatively small granite block requires an enormous amount of work and that it cannot be done too often for obvious economic reasons.

Therefore, models are required. We shall first examine two extreme models, the first one being totally deterministic and the second one being totally random.

3.3.2 Deterministic models

Historically, the first fracture networks which were devised, were deterministic and quite simple.

For instance, Warren and Root (1963) modeled a fractured porous rock as an idealized system made up of identical rectangular porous parallelepipeds, separated by an orthogonal network of infinite fractures. For obvious reasons in Fig. 3.7, this is called the *sugar box model*. Flow is assumed as simultaneously taking place in the fracture network and in the porous blocks with transfers between these two structures.

Fig. 3.7 The classical and deterministic sugar box model of Warren and Root (1963).

3.3.3 Random models

These models, to some extent, comprise the opposite of the previous ones.

In order to generate such models, one needs to make decisions with regard to a large number of parameters that can be listed as follows:

(1) The shape of the fracture; for instance, all the fractures of the network can be hexagonal or circular.
(2) The lateral extension; as already mentioned in Section 3.1, this extension can be characterized by $2R$ where R is the radius of the circumscribed circle to the fractures.
(3) The statistical properties of the positions of the fracture centers; for instance, one of the simplest assumptions is that they are randomly and uniformly distributed.
(4) The orientation of the normal n to the fracture; again the simplest assumption is that the normals are isotropically distributed over the unit sphere.
(5) The in-plane orientation angle ω which can be assumed to be isotropically distributed (see Fig. 3.4).
(6) Finally, one needs to define the fracture density ρ which is the number of fractures per unit volume. This density can be assumed to be constant or variable in space as mentioned in (3).

A simple example of such a random network was given in Fig. 3.1b. It represents a random network of hexagons which are uniformly distributed in space and isotropically oriented; the size of the hexagons is constant. Except for the shape, this is the situation which will be mainly studied in most of this book.

Of course, all the six characteristics listed above can be modified at will and the number of different cases for study is extremely large. The rest of this subsection displays examples where one or several of these five characteristics is modified.

First consider fractures with different shapes and various sizes. Rectangles with an aspect ratio equal to 4 are displayed in Fig. 3.8a; they are inscribed in circles of radius R; the probability density of the radii is assumed to be a power law with an exponent equal to 1.5 (see eqn 3.26); the ratio between the largest and the smallest values of R is equal to 10. A similar case is demonstrated for polydisperse hexagons in Fig. 3.8c.

It might be interesting to notice here that the probability density of R can be derived from measurements. For instance, if the fractures are assumed to be circular disks which are isotropically and uniformly distributed, the probability density of R can be deduced from the distributions of the chords $g(c)$ (cf. Section 3.2.1); two examples of such a procedure are provided by Berkowitz and Adler (1998) and Patriarche et al. (2007).

The fractures are not always uniformly distributed. For instance, the density ρ of the fractures generated by the digging of a gallery underground decreases with the distance to the gallery. Measurements show

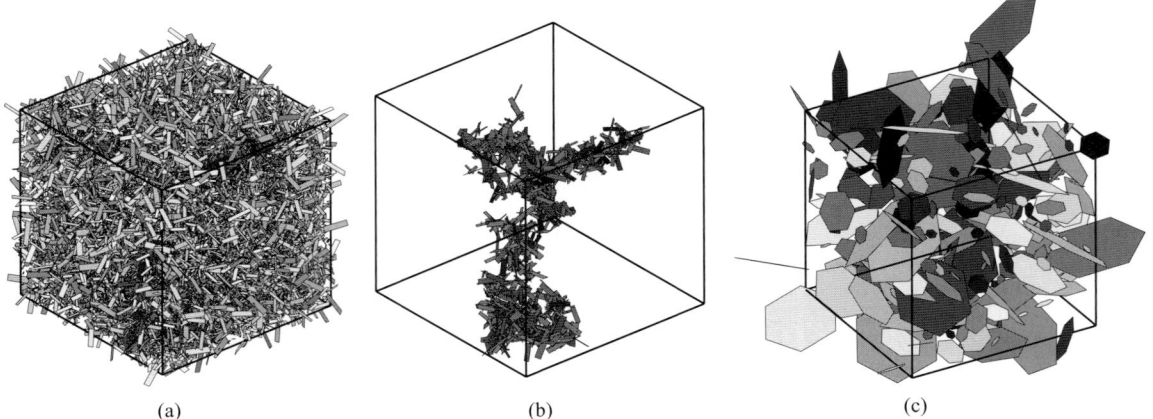

Fig. 3.8 Examples of polydisperse networks. R_M is the maximal radius of the circles circumscribed to the fractures. Rectangular fractures are shown in (a) and (b); (a) is the full network with $N_{fr} = 65\,900$ fractures in a $(16R_M)^3$ unit cell; (b) shows the corresponding spanning cluster. A polydisperse network of 300 hexagons in a $(4R_M)^3$ unit cell is shown in (c).

that ρ often follows an exponential law (Bossart *et al.*, 2002; Thovert *et al.*, 2011)

$$\rho(x, y, z) = \rho_o \exp(-z/\ell_d) \tag{3.2}$$

where ρ_o is the density at the wall and ℓ_d a characteristic decay length. Such a network is illustrated in Fig. 3.9a.

Anisotropic fracture networks are another important category. The standard distribution is the Fisher distribution (see Section 3.8); the normals to the fractures are more or less oriented around a specific direction called the pole, characterized by the unit vector $\boldsymbol{p}_F$. An example of an anisotropic network with $\boldsymbol{p}_F$ parallel to the z-axis is displayed in Fig. 3.9b.

As emphasized in Section 2.4.1 and illustrated in Fig. 2.5, real fractures are composed of contact zones of zero aperture and of open zones. This general situation can be schematized by fractures which are either open or closed, as illustrated in Fig. 3.9c.

Finally, *hierarchical* fracture networks are composed of fractures of various ages as displayed in Fig. 3.9d. There is a pre-existing family of fractures (thick lines); then, another seismic event creates another family (thin lines). Usually, the more recent fractures tend to cease when they encounter the older family, but of course the probability of stopping is not 1.

To conclude this short overview of various random generations of networks, note that non-uniformly distributed networks have not been considered in this list, apart for the one particular case (3.2). Note also that all the previous cases can be combined; to give a few examples, one can mix fractures of different shapes, add several families of anisotropic fractures, generate inhomogeneous and anisotropic networks and so on. It is clear that the number of combinations is almost unlimited.

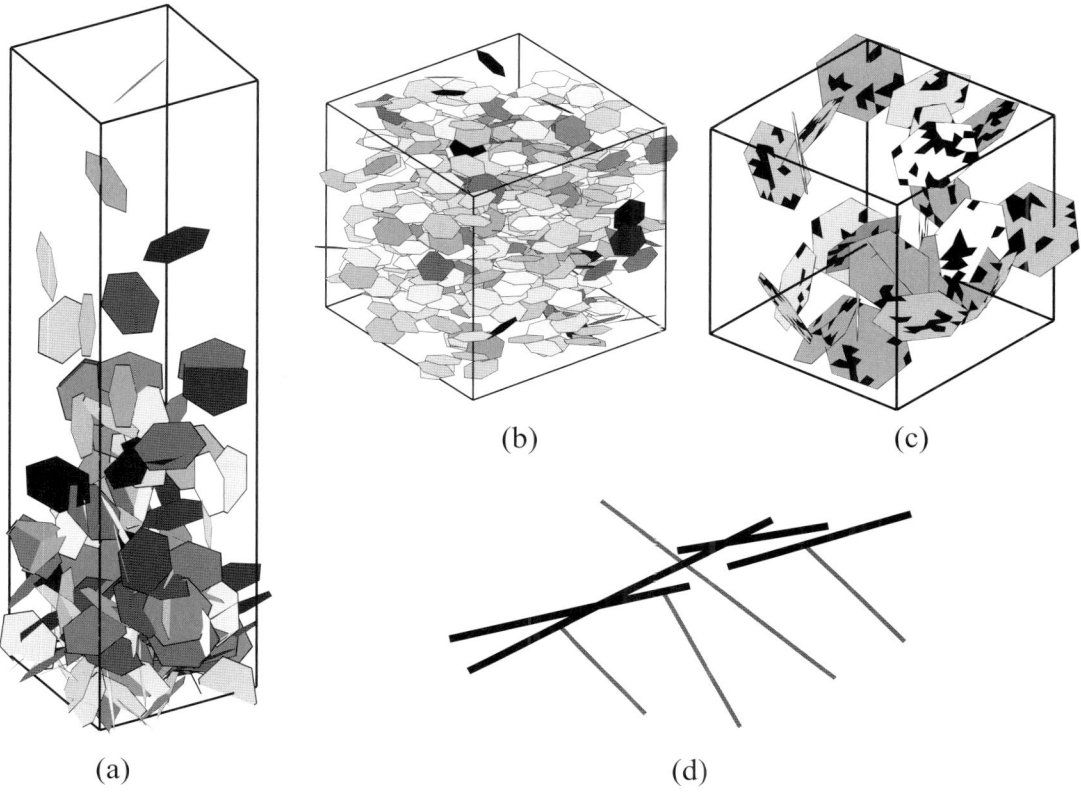

Fig. 3.9 Examples of: (a) monodisperse hexagons distributed according to an exponential density law in a $5 \times 5 \times 20R^3$ cell; (b) anisotropic network in a $(12R)^3$ cell composed of fractures approximately parallel to the xy-plane since the Fisher pole is parallel to the z-axis; (c) monodisperse hexagons in a $(5R)^3$ cell with open (white) and closed (black) zones; (d) a 2D hierarchical network with an old family of fractures (black thick lines) and a more recent one (grey thin lines).

3.3.4 Use of a library of data

It is often a good idea to use the data directly, especially when they are sufficiently numerous. Some examples of such use are provided in the rest of this subsection.

Let us once more consider the fracture measurements in the Baget watershed which are described in Section 3.2.2. The fracture density ρ is estimated by methods detailed in Section 3.5. The networks are generated in the following way. The measurements provide a library of 46 events, each one corresponding to a fracture which has been recorded. Then, one event is chosen at random with the help of a random number generator from this list of 46 events. A fracture is generated in the cell with the same orientation as the event which has been selected; the location of the center of the fracture is random and uniformly distributed over the cell. Then, other events are selected at random until the desired fracture density ρ is reached. All the fractures are of the same size.

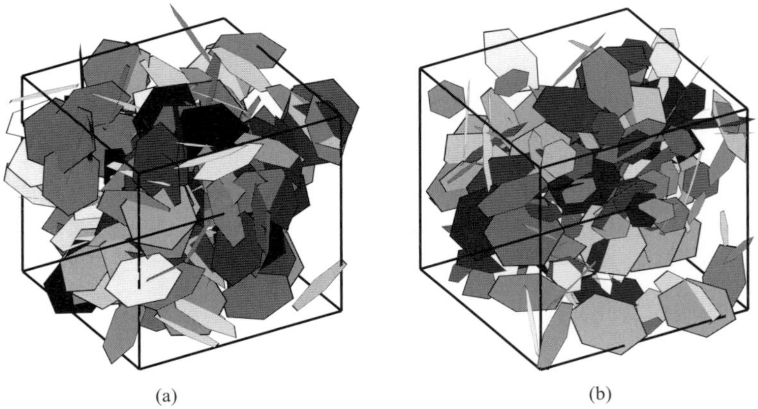

Fig. 3.10 Monodisperse (a) and bidisperse (b) samples simulated with the help of the data reported in Fig. 3.5. Data are for: $R = 50$ m (a), 25 and 50 m (b). The $(300 \text{ m})^3$ cube is the unit cell of an infinite spatially periodic medium.

Of course, this elementary rule can be made more complicated. Hexagons with two sizes $R = 25$ and 50 m can be generated. This is illustrated in Fig. 3.10 (see also Sisavath *et al.*, 2004).

Very often the data are site specific and they cannot be easily transposed from one site to another. For instance, Patriarche *et al.* (2007) distinguished two families of fractures based on trace lengths in the Roselend tunnel. The small fractures were isotropically distributed with a power law for the probability density of the fracture radii; the power law was deduced from the measurements. The large fractures had known orientations and they were supposed to have a single radius equal to 5 m. Fracture networks were generated by combining random models (the small fractures) and using a library (the large fractures).

3.3.5 Some concluding remarks

The generation of fracture networks depends crucially on the field observations which are communicated to the modeler. In practice, no field is strictly identical to another. This feature constitutes the major difficulty and advantages of such studies.

What should be done is an a priori study of generic cases, such as uniformly and isotropically distributed networks of monodisperse fractures, which is the simplest possible random geometry. In a sense, such networks are not meant to correspond to real ones, but to provide examples which can be fully worked out, analyzed and understood.

Most of the rest of this chapter is based on this first generic case, except for Section 3.8. The major properties of a fracture network are its percolating character and its permeability.

3.4 Statistical geometrical properties of fracture networks

When considering a fracture network, one needs to study its major geometrical properties, since permeability depends only on its geometry according to (1.4). The most important question regarding a network is whether it percolates or not, because this has a crucial influence on its permeability.

Another question is the determination of the number of *solid blocks* which are cut in the matrix by the network. Of course, when the fracture density ρ is small, fractures are isolated and there is only one infinite block which is the medium itself. When ρ increases, blocks are likely to become more and more numerous. The industrial requirements of these blocks depend on the application. In order to store nuclear waste in granite repositories, for example, blocks should be as large as possible. However, in an oil field, the oil is mostly in the porous medium from which it flows to the fractures and hopefully to the wells; in this case, the blocks should be as small as possible.

One should note that ρ for fracture networks plays the role of the probability P for the bond and site networks which were described in Section 2.5.

A final technical point is that most of these geometrical quantities are calculated for spatially periodic networks. A unit cell is generated and the medium is supposed to be the juxtaposition of an infinite number of unit cells as explained in Section 2.4.2. Because of this property, some fractures centered in the unit cell straddle the unit cell walls as can be seen in Figs 3.1b, 3.8–3.10.

3.4.1 A unifying concept: the excluded volume

The *excluded volume* V_{ex} of an object was defined by Balberg *et al.* (1984) as the volume surrounding it, in which the center of another object must be found in order for the two objects to intersect. This abstract definition can be easily understood by considering two spheres of radii R_1 and R_2 as shown in Fig. 3.11. Suppose that sphere 1 is maintained fixed; sphere 1 can be called the reference sphere; then, sphere 2 does not overlap sphere 1 when the center of sphere 2 lies outside sphere S_e which has the same center as sphere 1 and whose radius is equal to $R_1 + R_2$. The volume of sphere S_e is by definition the excluded volume V_{ex} for the two spheres

$$\begin{aligned} V_{ex} &= \frac{4\pi}{3}(R_1 + R_2)^3 \\ &= \frac{32\pi}{3}R^3 \quad \text{if } R_1 = R_2 = R \end{aligned} \quad (3.3)$$

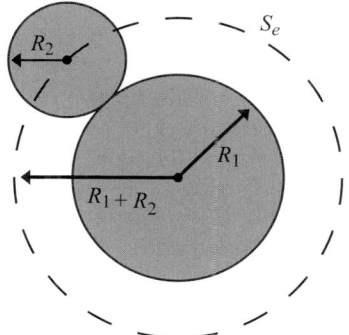

Fig. 3.11 The excluded volume of the two spheres is sphere S_e of radius $R_1 + R_2$ represented by the broken circle.

Note that it has been implicitly assumed that the density of sphere 2 is constant in space.

This notion which can be extended to ellipsoids for instance, is illustrated in Fig. 3.12. It is immediately apparent that the relative orientation of the two objects should be taken into account. An obvious way to cope with this difficulty is to calculate the excluded volume for a given relative orientation and then to average over all the orientations. It is also apparent that the spatial distribution of the objects is important. Therefore, assumptions about these orientations and this distribution are

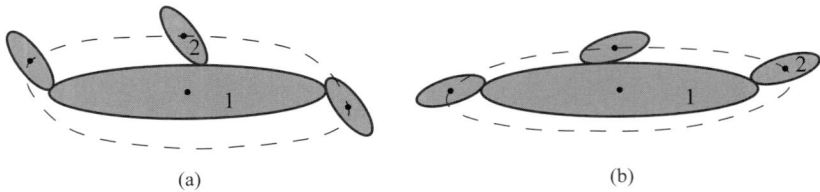

Fig. 3.12 Excluded volume of two ellipsoids. The reference ellipsoid 1 is kept fixed while ellipsoid 2 is moved around it with given orientations in (a) and (b). The excluded volume is tentatively indicated by the broken curve and it obviously depends on the orientation.

needed and in most of this chapter, the objects are assumed to be identical, isotropically oriented and uniformly distributed. *This case which is the reference one, is designated by the acronym I²OUD or IIOUD and the related quantities are indicated by the subscript r.* When the objects are not identical, the case is referred to as IOUD; for instance, a mixture of shapes may be IOUD. We shall often be very imprecise in our phrasing and say that the networks or the fractures are I²OUD or IIOUD.

The concept of excluded volume can be easily extended to two flat objects. For instance, an ellipse can be viewed as an ellipsoid with one semi-axis which tends towards zero. This concept can be extended to polygonal forms as well; this is illustrated in Fig. 3.13. Again the relative orientation of the two flat objects in space has to be taken into account.

A priori general calculations of excluded volumes seem to be very difficult. But, this apparent difficulty is an illusion when convex objects are considered.

A general expression for the excluded volume was established very early in the context of statistical mechanics by Isihara (1950) for IOUD objects. For two three-dimensional convex objects A and B with volumes V_A and V_B, areas A_A and A_B and mean radii of curvature R_A and R_B averaged over the contour orientation, he obtained the mutual exclusion volume

$$V_{ex,r,AB} = V_A + V_B + (A_A R_B + A_B R_A) \tag{3.4}$$

Note that this expression is symmetric in the sense that $V_{ex,r,AB} = V_{ex,r,BA}$. This expression can be averaged over the distributions of object shapes and sizes. For equal spheres, (3.3) is obtained. For flat convex

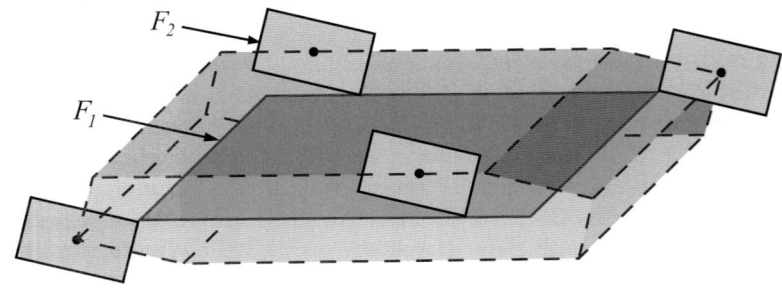

Fig. 3.13 Excluded volume of two rectangles F_1 and F_2 when one side of F_1 is parallel to F_2. The excluded volume of the two rectangles is tentatively indicated by the broken lines.

objects in space with perimeters P_A and P_B, Charlaix *et al.* (1984) reduced it to

$$V_{ex,r,AB} = \frac{1}{4}(A_A P_B + A_B P_A) \qquad (3.5a)$$

The derivation of this last formula is proposed in Exercise 3.1 which is highly recommended to the theoreticians! Its extension to convex polygons was detailed by Adler and Thovert (1999). When the two objects are identical with area A and perimeter P, this formula becomes

$$V_{ex,r} = \frac{1}{2}AP \qquad (3.5b)$$

Remember the hypotheses on which the relations (3.5) are based. They are valid for IOUD convex flat objects.

3.4.2 The dimensionless density ρ'

An important characteristics of a fracture network is its fracture density ρ which was defined in Section 3.3.3 as the number of fractures per unit volume. A *dimensionless density* ρ' can be introduced as the number of fractures per excluded volume

$$\rho' = \rho V_{ex} \qquad (3.6)$$

Since any fracture whose center is located in the excluded volume overlaps with the reference fracture, ρ' is obviously equal to the average number of intersections per fracture. Therefore, ρ' is a measure of the connectivity of the fracture network.

This notion of dimensionless density is crucial since it will be shown that many of the macroscopic properties of the fracture networks, such as the percolation threshold, permeability and so on, depend almost entirely on ρ', whatever the shape of the fracture.

3.4.3 The percolation threshold of identical convex fractures

The methodology described in the first part of this section is general and can be applied to any set of fractures, but the results (starting from eqn 3.10 and Fig. 3.15) are relative to fractures which are assumed to be isotropically oriented and uniformly distributed in addition to being identical and convex (i.e. I^2OUD).

The percolation threshold ρ_c is defined as being the fracture density above which the fracture network percolates; it is determined by numerical computations. ρ_c plays the role of the critical probability P_c introduced for bond and site networks in Section 2.5. When ρ is smaller than ρ_c, the fracture network does not percolate.

The numerical code is organized as follows. A first network is constructed by generating polygons of a given shape with a lateral extension R at random in a unit cell of size L with spatially periodic boundary conditions. The intersections of all the fractures are determined and an intersection table can be filled.

Then, the connected components of the fracture network are determined by a pseudo-diffusion algorithm which works as follows. Let us start with fracture 1; label it as fracture 1; look at the intersection table and give label 2 to all the fractures which intersect fracture 1; look at the intersection table of fractures with label 2 and give label 3 to all the fractures which intersect them; repeat this process until no new fracture is obtained. This set will be the connected component of fracture 1.

Remove the connected component of fracture 1 from the initial set of fractures. Choose a fracture at random and call it 2 for the sake of simplicity. As before determine the connected component of fracture 2.

Remove the connected component of fracture 2 from the previous set of fractures (therefore, without component 1) and repeat the process until there is no fracture left.

These components can be classified into percolating and non-percolating components. In a spatially periodic medium, a connected component is said to be percolating if it connects two opposite faces of the parallelepipedic unit cell, and if it contains two homologous fractures, i.e. two fractures with the same coordinates, modulo the spatial period $\boldsymbol{X}_{n,m}$ defined by eqn 2.25b. Components 1 and 2 in Fig. 3.14 provide examples of percolating and non-percolating components, respectively.

This series of operations is repeated for a sufficiently large number $\mathcal{N}$ of independent networks. Among these $\mathcal{N}$ networks, $\mathcal{N}_p$ percolate. Therefore, the probability of percolation $\Pi_L(\rho)$ of networks with a density ρ is equal to

$$\Pi_L(\rho) = \frac{\mathcal{N}_p}{\mathcal{N}} \tag{3.7a}$$

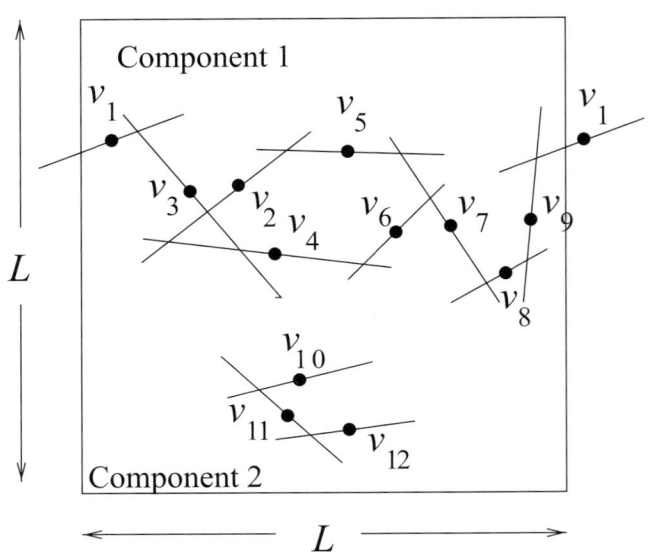

Fig. 3.14 The connected and percolating components in a two-dimensional fracture network; the square is the unit cell of a 2D spatially periodic medium. The fractures are the random segments denoted as v_i. The network consists of two connected components; 1 percolates through the cell, and 2 does not. Periodic conditions are used for component 1; v_1 straddles the unit cell wall; it is connected to v_3 in the left part of the unit cell and to v_9 in the right.

Since the networks are independently and identically distributed, one expects $\Pi_L(\rho)$ to be an error function (as the limit of the law of large numbers; cf. Papoulis, 1991)

$$\Pi_L(\rho) = \frac{1}{\sqrt{2\pi}\Delta_L} \int_{-\infty}^{\rho} \exp\left\{-\frac{(\xi - \rho_{Lc})^2}{2(\Delta_L)^2}\right\} d\xi \quad (3.7b)$$

where ρ_{Lc} and Δ_L are fit parameters. As discussed in Section 2.5, because of the finite-size effect, ρ_{Lc} and Δ_L should be evaluated for several values of L; this effect was illustrated in Fig. 2.11. The asymptotic value ρ_c of ρ_{Lc} for infinite systems can be derived from the two scaling relations (Stauffer and Aharony, 1994)

$$\rho_{Lc} - \rho_c \propto L^{-1/\nu} \qquad \Delta_L \propto L^{-1/\nu} \quad (3.8)$$

The plots of $\rho_{Lc}(\Delta_L)$ are extrapolated for $\Delta_L \to 0$ to determine ρ_c.

This set of operations can be repeated for a number of shapes and Huseby *et al.* (1997) summarized their results in Fig. 3.15a where the diameter $2R$ is the length unit for the definition of ρ. These data can be expressed in terms of the dimensionless percolation threshold according to the definition (3.6)

$$\rho'_c = V_{ex}\rho_c \quad (3.9)$$

Of course, in this formula, the excluded volume is calculated for each shape by using eqn 3.5b. When this is done, a remarkable property appears for I²OUD fractures. The dimensionless percolation threshold appears to be independent of the fracture shape (cf. Fig. 3.15b)

$$\rho'_{c,r} \approx 2.28 \pm 0.07 \quad (3.10)$$

This result is very important since it demonstrates the usefulness of the concepts of excluded volume and dimensionless density. Moreover, similar properties will be shown for permeabilities of fracture networks and of fractured porous media in Chapters 5 and 6, respectively. However, it should also be noticed that this first result is not precise and that the more precise relation (3.11) was derived several years later.

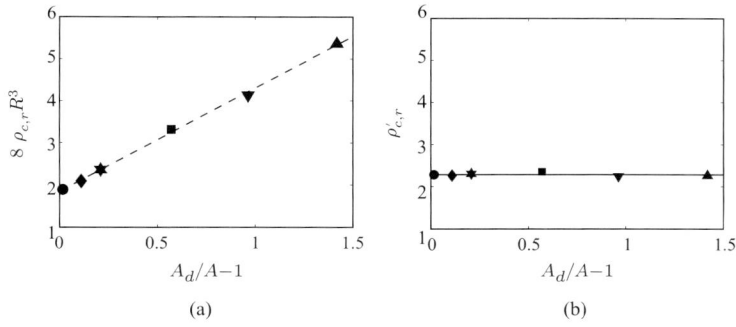

Fig. 3.15 (a) The percolation thresholds $8\rho_{c,r}R^3$ as functions of $(A_d/A - 1)$ where A_d is the area of the circumscribed disk. Data are for I²OUD fractures: 20-gones (●), octogons (♦), hexagons (★), squares (■), 2-rectangles (▼), triangles (▲). (b) The dimensionless percolation threshold $\rho'_{c,r}$ for the same shapes.

Table 3.1 Dimensionless parameters associated with disks and with the investigated fracture shapes. R is the radius of the circle circumscribed to the fracture.

	Disks	Hexagons	Squares	Triangles	6-rectangles
P/R	2π	6	$4\sqrt{2}$	$3\sqrt{3}$	$28/\sqrt{37}$
A/R^2	π	$3\sqrt{3}/2$	2	$3\sqrt{3}/4$	$24/37$
$V_{ex,r}/R^3$	π^2	$9\sqrt{3}/2$	$4\sqrt{2}$	$27/8$	$336\sqrt{37}/1369$
η_p	$2/\pi$	$2/3$	$\sqrt{2}/2$	$4\sqrt{3}/9$	$\sqrt{37}/7$

One could be tempted to provide a simple interpretation to (3.10). One could say that this result is very normal since if a network percolates, one needs one fracture to reach a given fracture and another fracture to leave this given fracture. Therefore, ρ'_c should be at least equal to 2; so far the argument is valid. But, any justification by some hand-waving argument that about 10% of the fractures end nowhere, which explains the slightly larger than 2 result of 2.28, is fallacious since for randomly oriented sticks in two dimensions, $\rho'_{c,r}$ is equal to 3.6 which is much more than 2!

These first results were improved by Mourzenko et al. (2005). $\rho'_{c,r}$ was found to be equal to 2.31 for hexagons which is the most extensively studied shape, and $\rho'_{c,r}$ is nearly invariant except for very slender fracture shapes. All the data could be summarized by the formula

$$\rho'_{c,r} = 2.41 \left[1 - 4\left(\eta_p - \frac{2}{\pi}\right)^2\right] \pm 0.1, \quad \text{with} \quad \eta_p = \frac{4R}{P} \qquad (3.11)$$

where the shape factor η_p is given in Table 3.1 for the five fracture shapes. This yields $\rho'_{c,r} = 2.36$, 2.24 and 1.90 for squares, triangles and 6-rectangles, respectively. For the sake of clarity, an n-rectangle is a rectangle with an aspect ratio which is equal to n. For very elongated rectangles, η_p tends towards 1. Therefore, η_p varies between $2/\pi$ and 1.

$\rho'_{c,r}$ decreases from 2.41 to 1.14 when the shape changes from a disk to an infinitely elongated rectangle.

Exercises 3.2–3.4 are direct applications of the concepts and results presented in this section.

3.4.4 Solid blocks

When the fracture density ρ is large enough, the network can cut through the solid matrix solid blocks whose number per unit volume is called *solid block density* and is called ρ_b. This simple concept is illustrated in Fig. 3.16a.

The fracture networks are generated as in Section 3.4.3. The physical idea which is behind the algorithm detailed by Huseby et al. (1997) is relatively simple. A solid block is limited by several plane faces which are located within the construction polygons; to recognize such a block requires at least two steps: namely a recognition of the plane faces and the identification of faces which are neighbors one to another. A simple

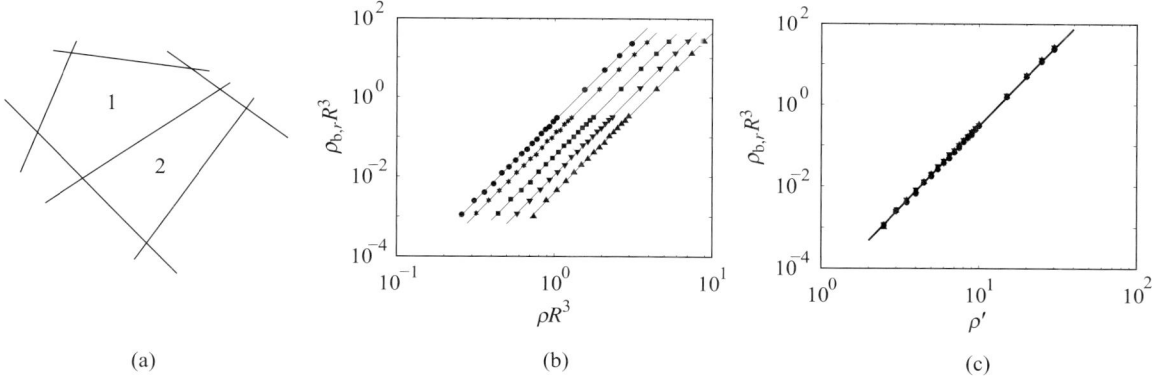

Fig. 3.16 Solid blocks cut in the solid matrix by the fracture network. The concept of solid block is illustrated in two dimensions in (a); the fractures symbolized by the segments cut in the plane two blocks denoted by 1 and 2. (b) The dimensionless solid block density $\rho_{b,r}R^3$ as a function of the dimensional fracture density ρ for various I²OUD shapes. (c) The dimensionless solid block density $\rho_{b,r}R^3$ as a function of the dimensionless fracture density ρ' Conventions are the same as in Fig. 3.15.

example is provided by a cube with six faces; each face is the neighbor of four others. Suppose that the faces are translated far away one from another; since the faces do not intersect, they are not neighbors and thus they do not bound a solid cube any longer. A more subtle example is the initial cube, but with one face suppressed; here again, the remaining five faces do not bound a solid block. All possible cases are addressed by the algorithm, and the interrelations of the fracture faces are summarized by a graph in the sense of the graph theory; the fracture faces are the vertices of the graph which are linked by an edge when they intersect.

Results are displayed in Fig. 3.16b in dimensional terms for various I²OUD shapes. The points gather well around straight lines for each fracture shape, i.e. around power laws with exponents between 4.0 and 4.1. An exponent equal to or greater than 4 can be rationalized by considering the fact that the formation of a polyhedron requires the intersection of at least four fractures. However, the lines in Fig. 3.16b spread vertically over nearly two orders of magnitude, whereas the same data plotted as functions of ρ' collapse onto a single line

$$\rho_{b,r}R^3 = 3 \cdot 10^{-5} \, \rho'^4 \qquad (3.12)$$

The data for all regular polygons and for rectangles with an aspect ratio of 2 are represented with a root-mean-square error of 6.7% in the investigated range. Note that the data in Fig. 3.16 were obtained recently with much larger sample sizes than in Huseby et al. (1997) which explains the differences in the results.

Shape effects are illustrated by Exercise 3.5.

3.4.5 Concluding remarks

It has been shown in the previous examples that ρ' is a crucial quantity whose knowledge is a key to basic properties of fracture networks.

3.5 Estimation of the dimensionless density from line data

Therefore, the next question is how to extract this information from field measurements. As mentioned in Section 3.1, fractures are only visible on outcrops and measurements can be performed along lines or over surfaces. It is the purpose of the next two sections to show how such information can be derived by elementary means.

Line data are obviously the easiest ones to obtain. For instance, the number of intersections N_I on a line of length L with traces, can be measured.

Consider a line of length L and a population of convex fractures which are isotropically and uniformly distributed in space. The area and the perimeter of the fractures are denoted by A and P, respectively. A priori, this problem illustrated in Fig. 3.17a is complex and can only be handled analytically. But, there is an easy way to obtain the result by using the concept of excluded volume again.

Simply transform the line into an elongated rectangle of length L and width ℓ as shown in Fig. 3.17b. Then, according to (3.5a), the excluded volume $V_{ex,r\ell}$ between the rectangle (inscribed in a circle of radius R) and the IOUD fractures is given by

$$V_{ex,r\ell} = \frac{1}{4}[\ell L \times P + 2(\ell + L)A] \qquad (3.13a)$$

Equivalently,

$$V_{ex,r\ell} = \frac{1}{2}AL\left[1 + \frac{\ell P}{2A} + \frac{\ell}{L}\right] \qquad (3.13b)$$

Now, let ℓ tend towards 0 to obtain

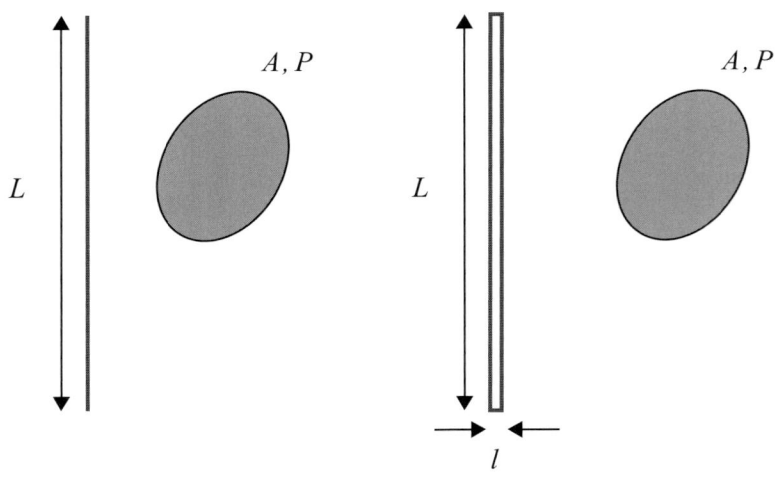

Fig. 3.17 Intersections between a line and a fracture. (a) Initial situation. (b) Introduction of an elongated rectangle in order to use the excluded volume between the fractures and the rectangle.

$$V_{ex,r\ell} = \frac{1}{2}AL \qquad (3.13c)$$

Since the fracture density is ρ, the number of intersections between the fractures and the line of length L is

$$N_{I,r} = \rho V_{ex,r\ell} = \frac{1}{2}\rho AL \qquad (3.14a)$$

Alternative expressions for the number of intersections per unit length $n_{I,r}$ can be proposed

$$n_{I,r} = \frac{N_{I,r}}{L} = \frac{1}{2}\rho A \qquad (3.14b)$$

ρ can be replaced by ρ' by using (3.6). Since the excluded volume is given by (3.5b), one obtains a dimensionless relation

$$\rho' = Pn_{I,r} \qquad (3.14c)$$

For I^{2}OUD disks of radius R, this last formula becomes

$$\rho' = 2\pi R n_{I,r} \qquad (3.14d)$$

Another useful formula is the mean spacing $\bar{s}$ between two successive intersections. Equation 3.14c implies

$$\bar{s} = \frac{P}{\rho'} \qquad (3.14e)$$

The previous development illustrates the power of geometrical reasoning. Another example of this power is the following extension to anisotropic fracture distributions. Fig. 3.18a shows a line of length L parallel to the unit vector $\boldsymbol{p}$ and a family of fractures of area A and perimeter P which are normal to the unit vector $\boldsymbol{n}$.

There is an easy way to calculate the number of intersections between this line and the family of fractures. Consider the cylinder whose axis is the line L and whose basis is one fracture (see Fig. 3.18b). The volume V of this cylinder is thus

Fig. 3.18 Intersections between a line and a family of fractures with a given orientation. (a) Initial situation. (b) The cylinder of volume V.

$$V = LA|\boldsymbol{p} \cdot \boldsymbol{n}| \tag{3.15}$$

The fractures whose centers belong to this volume intersect the line. Since the fracture density is ρ, N_I can be expressed as

$$N_I = \rho V = \rho LA|\boldsymbol{p} \cdot \boldsymbol{n}| \tag{3.16a}$$

The number of intersections per unit length n_I is expressed as

$$n_I = \rho A|\boldsymbol{p} \cdot \boldsymbol{n}| \tag{3.16b}$$

Therefore,

$$\rho' = \frac{n_I}{2} \frac{P}{|\boldsymbol{p} \cdot \boldsymbol{n}|} \tag{3.16c}$$

It should be noted that in the formulae (3.14c) and (3.16c), one needs to know the perimeter P of the fracture. In other words, one needs to have an order of magnitude of the lateral extension of the fracture. This remark has to be kept in mind.

These last formulae can be extended to families of fractures of various characteristics such as shapes, lateral extensions and orientations. The total number of intersections is the sum of the intersections corresponding to each family. Therefore,

$$n_I = \sum_f \rho_f A_f |\boldsymbol{p} \cdot \boldsymbol{n}_f| \tag{3.17}$$

where the subscript f denotes a given family. This last relation is the starting point of Exercise 3.6.

3.6 Estimation of the dimensionless density from surface data

The number of traces present on an outcrop can be easily measured, at least in principle.

For IOUD fractures, one can calculate the number of traces on a plane Π by once more using the concept of excluded volume. This reasoning is illustrated in Fig. 3.19. Consider an imaginary convex surface S (of area A_S and perimeter P_S) which is drawn on the plane Π. The excluded volume $V_{ex,rS}$ between the surface S and the fractures is once again given by eqn 3.5a. Therefore,

$$V_{ex,rS} = \frac{1}{4}(A_S P + A P_S) \tag{3.18}$$

The number of intersections between the fractures and the surface S can be expressed as

$$\mathcal{N} = \rho V_{ex,rS} = \frac{\rho}{4}(A_S P + A P_S) \tag{3.19a}$$

The number of intersections per unit surface Σ_t, which was introduced in Section 3.2.1, is expressed as

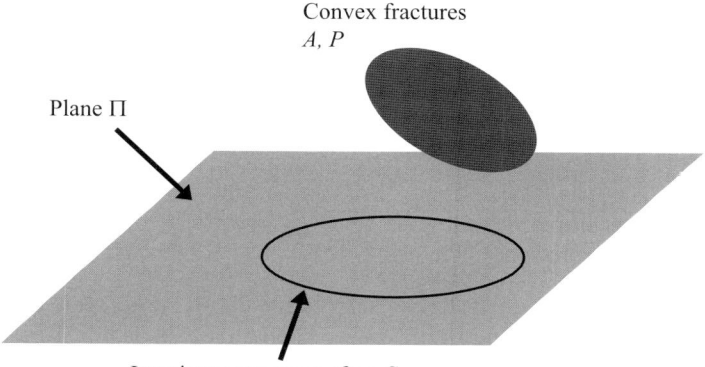

Fig. 3.19 Intersections between a plane Π and a family of fractures.

$$\Sigma_{t,r} = \frac{\rho}{4}\left(P + A\frac{P_S}{A_S}\right) \qquad (3.19b)$$

When the surface A_S becomes very large, the last term in the right-hand side goes to zero. Therefore,

$$\Sigma_{t,r} = \frac{\rho P}{4} \qquad (3.19c)$$

or in dimensionless terms

$$\rho' = 2A\Sigma_{t,r} \qquad (3.19d)$$

a relation which is valid whatever the fracture shape if it is convex.

As before, these relations can be generalized to families with known orientations. Now, note that it is the fracture area which is required in order to obtain the dimensionless density, in contrast with eqn 3.14c where it was the perimeter.

A very important application of the concepts and relations presented in the previous section and in this one is proposed in Exercise 3.7. It is very similar to a real problem in that not all the data are available and that hypotheses should be made in order to reach some conclusions.

3.7 Stereological relations for convex fractures

Stereology is the science which consists of relating observations made in one or two dimensions to the real world which exists in three dimensions. Some stereology has already been used in Sections 3.5 and 3.6 since the volumetric fracture density was related to the number of intersections of the network with a line or a plane.

Additional relations can be established and it is the purpose of this section to present them and to show how they can be used practically.

3.7.1 The direct stereological relations

Let us first recall some notations which were defined in Section 3.2.1. On an outcrop, one can measure the average density of intersections along a line n_I, the surface density of traces Σ_t, the average trace length $\langle c \rangle$ and the surface density of intersections between fracture traces Σ_p.

Thovert and Adler (2004) determined a series of relations which are valid for networks of convex fractures which are uniformly distributed in space, but not necessarily isotropic. These relations are contained in Table 3.2. Note that *subvertical* simply means that the fractures are approximately orthogonal to the horizontal plane; of course, these relations also apply when the fractures are perpendicular to a vertical observation plane.

In Table 3.2, $\boldsymbol{p}$ is the unit vector which is parallel to the observation line in the expressions for n_I when $\boldsymbol{n}$ is the unit vector which is normal to the fractures. In the second and third lines of the table, α is the angle between $\boldsymbol{n}$ and the normal $\boldsymbol{\nu}_\Pi$ to the observation plane Π. If Π is horizontal, α corresponds to the dip angle θ (cf. Figs 3.4a, b). Moreover, the two following quantities are averages over any two fractures labeled as 1 and 2 in the network

$$\mathcal{A}_{12} = \langle A_1 A_2 | \boldsymbol{\nu}_\Pi \cdot (\boldsymbol{n}_1 \times \boldsymbol{n}_2) | \rangle, \quad \mathcal{B}_{12} = \langle A_1 A_2 | \sin \beta_{12} | \rangle \qquad (3.20)$$

where β_{12} is the angle between the normals $\boldsymbol{n}_1$ and $\boldsymbol{n}_2$ to the two fractures 1 and 2.

$\mathcal{C}$ is defined as the total trace length per unit surface. Therefore, it verifies

$$\mathcal{C} = \Sigma_t \langle c \rangle \qquad (3.21)$$

Its value can be readily derived from this relation.

The relations of Table 3.2 present a number of remarkable features. First of all, they are remarkably simple for IOUD fractures. The anisotropic case is basically no more difficult, but it requires relevant information on the fracture orientations. Second, since all average quantities are additive, these relations are valid for mixtures of sizes and also mixtures of shapes, provided that all the fractures are convex. Therefore, these relations are quite general and important.

Table 3.2 The major relations for the various kinds of networks. The fractures may be of different sizes and shapes, but they are always uniformly distributed (UD).

	Isotropic 3D	Anisotropic 3D	Subvertical isotropic	Subvertical anisotropic
$\langle n_I \rangle$	$\frac{1}{2}\rho\langle A \rangle$	$\rho\langle A|\boldsymbol{p}\cdot\boldsymbol{n}|\rangle$	$\frac{2}{\pi}\rho\langle A \rangle$	$\rho\langle A|\boldsymbol{p}\cdot\boldsymbol{n}|\rangle$
Σ_t	$\frac{1}{4}\rho\langle P \rangle$	$\frac{\rho}{\pi}\langle|\sin\alpha|P\rangle$	$\frac{\rho}{\pi}\langle P \rangle$	$\frac{\rho}{\pi}\langle P \rangle$
$\langle c \rangle$	$\pi\frac{\langle A \rangle}{\langle P \rangle}$	$\pi\frac{\langle A|\sin\alpha|\rangle}{\langle P|\sin\alpha|\rangle}$	$\pi\frac{\langle A \rangle}{\langle P \rangle}$	$\pi\frac{\langle A \rangle}{\langle P \rangle}$
$\mathcal{C}$	$\frac{\pi}{4}\rho\langle A \rangle$	$\rho\langle A|\sin\alpha|\rangle$	$\rho\langle A \rangle$	$\rho\langle A \rangle$
Σ_p	$\frac{\pi}{16}\rho^2\langle A \rangle^2$	$\frac{1}{2}\rho^2 \mathcal{A}_{12}$	$\frac{1}{\pi}\rho^2\langle A \rangle^2$	$\frac{\rho^2}{2}\mathcal{B}_{12}$

3.7.2 The inverse stereological relations

Let us concentrate on the first two columns in Table 3.2 which correspond to the reference IOUD case. Column 1 corresponds to the quantities which can be measured on the outcrop, and column 2 contains crucial three-dimensional information such as fracture density ρ, average fracture surface $\langle A \rangle$ and average perimeter $\langle P \rangle$. Therefore, we want to invert these relations, i.e. to express ρ, $\langle A \rangle$ and $\langle P \rangle$ as functions of the measurable quantities.

Unfortunately, this inversion does not yield anything. When one tries to express one of these three-dimensional quantities as a function of the measurable ones, one always obtains $1 = 1$ which is certainly not wrong, but is not very useful! Actually, only two of the three quantities ρ, $\langle A \rangle$ and $\langle P \rangle$ can be simultaneously derived from the average measured data.

One way to obtain information is to define a shape factor such as

$$\eta = \frac{\langle A \rangle}{\langle P \rangle^2} \tag{3.22}$$

Of course, this choice is not unique and η_p described in (3.11) could be chosen. Then, ρ, $\langle A \rangle$ and $\langle P \rangle$ can be expressed as

$$\langle P \rangle = \frac{\langle c \rangle}{\pi \eta}, \quad \langle A \rangle = \frac{\langle c \rangle^2}{\pi^2 \eta}, \quad \rho = \frac{\pi^2 \eta \Sigma_t}{\langle \sin \alpha \rangle \langle c \rangle} \tag{3.23a}$$

For isotropic networks, ρ can be simplified as

$$\rho = 4\pi\eta \frac{\Sigma_t}{\langle c \rangle} \tag{3.23b}$$

Then, the dimensionless density ρ' can be derived if the average excluded volume is known. One way to derive this last quantity would be to average directly (3.5a) and to assume that A and P are proportional to R^2 and R, respectively. Since the two fractures are independent, one has

$$\rho' \propto \frac{\rho}{2} \langle R^2 \rangle \langle R \rangle \tag{3.24}$$

where R is the lateral extension. But, this formula should be considered with care. More details on this question are given in Section 3.8.1.

Therefore, the situation is somewhat frustrating since all the major quantities depend on ρ' which cannot be determined unless some additional hypotheses are made. Well that's is life, and just another illustration of the American saying "There is no free lunch" whose translation in British English is "Nothing in life is free".

3.7.3 Consistency relations

There is another way to use the relations of Table 3.2 which in a sense is in complete opposition with the previous one. Let us eliminate the three-dimensional unknown quantities ρ, $\langle A \rangle$ and $\langle P \rangle$ to obtain the following three ratios which should be equal to one

$$\zeta_1 = \frac{\pi}{2} \frac{\langle n_I \rangle}{\Sigma_t \langle c \rangle}, \quad \zeta_2 = \frac{\pi \Sigma_p}{\Sigma_t^2 \langle c \rangle^2}, \quad \zeta_3 = \frac{\pi}{4} \frac{\langle n_I \rangle^2}{\Sigma_p} \qquad (3.25)$$

The third relation is derived by eliminating $\Sigma_t \langle c \rangle$ between ζ_1 and ζ_2. These relations provide consistency relations between the data. If ζ_1, ζ_2 or ζ_3 are very different from 1, the data should be carefully reconsidered. Most likely, it means that some of the requirements are not fulfilled; the network may not be IOUD and/or without any correlation between fracture positions and orientations.

Note that ζ_1 is insensitive to the spatial organization, and that this is not valid for ζ_2 and ζ_3 which depend on trace intersections. This has been applied to a real fracture field by Thovert and Adler (2004). An application is proposed in Exercise 3.8.

3.8 Extensions

The content of this section can be skipped in a first reading. It presents some recent developments which are not contained in our previous book and which go beyond the simplest possible case of identical fractures which are isotropically and uniformly distributed. These three restrictions will be progressively removed. One important feature of this section is that it is not really as self-contained as the others. Only summaries with plausibility arguments are given, and the technical details as well as the precise derivations can be found in the references.

First, fracture networks with power-law size distributions are studied in Section 3.8.1; the size distribution is defined and the percolation threshold can be derived from (3.32) and (3.34).

Then, anisotropic networks of convex fractures are defined as following a Fisher distribution for their orientations; various stereological relations are given in Section 3.8.2; the percolation threshold of such networks is given in Section 3.8.3 (cf. eqn 3.44).

Stereological relations for anisotropic and heterogeneous networks of convex fractures are gathered in Section 3.8.4.

Finally, the percolation threshold of networks of heterogeneous fractures (i.e. with a binary distribution of apertures) is studied in Section 3.8.5.

3.8.1 Percolation of fracture networks with power law size distributions

Mourzenko et al. (2005) considered three-dimensional networks made up of IOUD fractures with plane polygonal shapes. These polygons may be regular or not, but all their vertices are supposed to lie on a circumscribed circle, whose radius R provides a measure of their size. In agreement with many observations of fractured rocks which were reviewed by Adler and Thovert (1999), the statistical distribution of the fracture sizes is supposed to be described by a power law

$$n(R) = \alpha R^{-a} \qquad (3.26)$$

where $n(R)\mathrm{d}R$ is the probability of fracture radii in the range $[R, R+\mathrm{d}R]$; α is a normalization coefficient, and the exponent a ranges between 1 and 5. In practice, R may vary over a large interval which can span five orders of magnitude, from size R_m of the microcracks to size R_M of the largest fractures in the system. The normalization condition implies that α verifies

$$\alpha = \frac{a-1}{R_\mathrm{m}^{1-a} - R_\mathrm{M}^{1-a}} \quad (a \neq 1) \tag{3.27a}$$

$$\alpha = \frac{1}{\ln R_\mathrm{M} - \ln R_\mathrm{m}} \quad (a = 1) \tag{3.27b}$$

According to a recent synthesis on the real distribution of fracture trace lengths in a plane intersecting a three-dimensional fracture network by Bonnet et al. (2001), it is reasonable to assume that a varies between 1.8 and 4.5 with a maximum likelihood of around 3.0. Examples of polydisperse networks were given in Fig. 3.8.

The percolation properties of such networks are investigated in a finite size L^3 cubic domain. Hence, two main cases can be distinguished, when R_M is significantly smaller or larger than the domain size L; in addition, a transition regime occurs when R_M and L are of comparable orders of magnitude. In all cases, R_m is supposed to be much smaller than L. The rest of this section is focused on the first situation $R_\mathrm{M} \ll L$. An example of a percolating cluster is shown in Fig. 3.8b; it is of a much smaller size than the full network displayed in (a).

The first task to be performed, and probably the most important one, is the derivation of an adequate definition of the network density. With this in mind, we introduce the volumetric number density of fracture per fracture size $F(R)$

$$F(R) = \rho \, n(R) \tag{3.28}$$

where $F(R)\mathrm{d}R$ is the number of fractures with a radius in the range $[R, R+\mathrm{d}R]$ per unit volume. Here, ρ is a constant independent of R.

The volumetric moments M_p of the fracture radii are defined as

$$M_p = \int_{R_\mathrm{m}}^{R_\mathrm{M}} R^p \, F(R) \mathrm{d}R \tag{3.29}$$

They can be expressed as

$$M_p = \rho\alpha \, \frac{R_\mathrm{M}^{p+1-a} - R_\mathrm{m}^{p+1-a}}{p+1-a} \quad (a \neq p+1) \tag{3.30a}$$

$$M_p = \rho\alpha \, (\ln R_\mathrm{M} - \ln R_\mathrm{m}) \quad (a = p+1) \tag{3.30b}$$

Hence, the moments of the radii $\langle R^p \rangle = M_p/M_0$ are given by

$$\langle R^p \rangle = \frac{1-a}{p+1-a} \, \frac{R_\mathrm{M}^{p+1-a} - R_\mathrm{m}^{p+1-a}}{R_\mathrm{M}^{1-a} - R_\mathrm{m}^{1-a}} \quad (a \neq 1, p+1) \tag{3.31}$$

The value of $\langle R^p \rangle$ for $a=1$ or $p+1$ can be obtained by combining eqns 3.30a and 3.30b.

Next, an extension of the definition of the dimensionless density is needed. This can be done as follows. The generalized dimensionless density is defined as

$$\rho'_3 = \rho \, v_{\mathrm{ex}} \, \langle R^3 \rangle \qquad (3.32\mathrm{a})$$

where

$$v_{\mathrm{ex}} = \frac{1}{2} \left\langle \frac{A}{R^2} \right\rangle \left\langle \frac{P}{R} \right\rangle \qquad (3.32\mathrm{b})$$

The brackets $\langle \ \rangle$ denote the statistical moments of R weighted by $n(R)$. Of course, ρ'_3 coincides with ρ' for networks with fractures of a single size and of a single shape.

The percolation of polydisperse networks of convex fractures was studied in terms of ρ'_3 in Mourzenko et al. (2005). The percolation threshold ρ'_{3c} is expected to depend on size L of the cell, on the exponent a, on the ratio between the two extreme radii $R'_m = R_m / R_M$ and on the shape of the fractures. To cut a long story short, only Fig. 3.20 is given and discussed.

Two types of quantities are reported in this figure. The black symbols correspond to $\rho_c \langle R^3 \rangle$ and we are going to consider them first. The most striking feature is that this quantity does not depend on the ratio R'_m; the values are almost perfectly constant as functions of R'_m for various shapes and various combinations of shapes. This is already a remarkable achievement which means that the right unit volume is $\langle R^3 \rangle$ in contradiction with the heuristic formula (3.24).

It should be noted that the mean number of intersections per fracture is provided by the following formula which is an extension of (3.24)

$$\rho' = \rho v_{\mathrm{ex}} \langle R^2 \rangle \langle R \rangle \qquad (3.33)$$

However, in polydisperse networks, connections between large fractures contribute more efficiently to percolation than connections between small ones. Large fractures are more heavily weighted in eqn 3.32a than in eqn 3.33 which explains that ρ'_3, rather than ρ', controls the percolation properties.

There is more to come! When the dimensionless excluded volume v_{ex} is taken into account and when ρ'_{3c} is plotted as a function of R'_m, all the previous data are gathered into a single horizontal curve which corresponds to a single critical value, with only a slight residual dependence on the fracture shape, which is well described by (3.11)

$$\rho'_{3c} = \rho'_{c,r} = 2.41 \left[1 - 4 \left(\eta_p - \frac{2}{\pi} \right)^2 \right] \pm 0.1 \qquad (3.34)$$

There is some difference between the historical value (3.10) which was obtained in 1997 and a recent result (3.11) in 2005. This is partly due to the fact that the 2005 results were obtained with larger statistical sets of samples of larger sizes. In addition, (3.11) applies to monodisperse and polydisperse networks with broad size distributions. The results obtained in 1997 for monodisperse networks are in the lower part of the ± 0.1 uncertainty interval in (3.11).

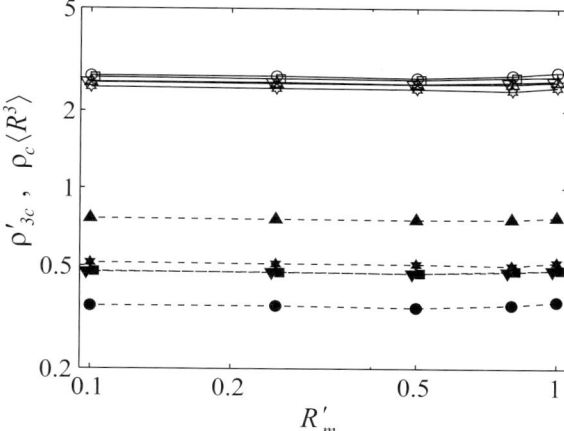

Fig. 3.20 The percolation thresholds ρ'_{3c} (open symbols, solid lines) and $\rho_c \langle R^3 \rangle$ (black symbols, broken lines) for non-periodic networks with $L = 6R_M$ and $a = 1.5$ for regular hexagons (○), squares (□), triangles (△), mixture of hexagons and triangles, 50%-50% (▽), and mixture of hexagons and rectangles with aspect ratio 4, 50%-50% (✡).

Finally, one should mention that the relationships between three-dimensional fracture networks consisting of polydisperse disks and the corresponding two-dimensional trace maps were systematically analyzed by Berkowitz and Adler (1998). The inverse problem, namely deriving the disk distribution from the trace distribution, was solved assuming that the disks are located and oriented randomly. These results were applied to a variety of synthetically generated data, and several sets of field data were considered.

Exercise 3.9 is a direct application of the results proposed in this section.

3.8.2 Stereological relations for anisotropic networks of convex fractures

Mourzenko *et al.* (2011a) considered anisotropic networks of plane fractures with convex contours of area A and perimeter P. Each fracture can be located in space when the position of its center $\boldsymbol{x}$, the normal $\boldsymbol{n}$ to the fracture and the angle ω which characterizes the fracture orientation within its plane relative to any arbitrary origin, are known. The components of $\boldsymbol{n}$ can be expressed in the standard polar coordinates θ and ϕ. All these notations are illustrated in Figs 3.4b,c where the Cartesian coordinates x, y and z are displayed. The intervals of variations of θ and ϕ are restricted to the upper hemisphere $\Omega/2$ of the unit sphere Ω, i.e.

$$0 < \phi \leq 2\pi, \quad 0 \leq \theta \leq \frac{\pi}{2} \qquad (3.35)$$

The fracture density ρ is constant in space; therefore, the fractures are uniformly distributed (i.e. UD).

The simplest probability density function $f(\theta, \phi)$ for the unit vector $\boldsymbol{n}$ is the uniform distribution which corresponds to an isotropic orientation of the fractures. It is expressed as

$$f_i(\theta, \phi) = \frac{\sin \theta}{2\pi} \quad \text{for} \quad \Omega/2 \qquad (3.36)$$

where the subscript i stands for isotropic.

The second most classical probability density function (PDF) is the Fisher distribution which corresponds to a random walk over the sphere (Mardia, 1972). Let (θ_0, ϕ_0) be the initial polar coordinates; assuming that the resulting distribution is rotationally symmetrical around the initial direction, the Fisher distribution is deduced as ($0 < \theta \leq \pi/2$, $0 < \phi \leq 2\pi$)

$$f_F(\theta, \phi) = \frac{\kappa}{2\pi \sinh \kappa} \sin \theta \; \cosh[\kappa\{\cos\theta_0 \cos\theta + \sin\theta_0 \sin\theta \cos(\phi - \phi_0)\}] \quad (3.37a)$$

The direction (θ_0, ϕ_0) is called the Fisher pole and denoted by $\boldsymbol{p}_F$. In the particular case where θ_0 is equal to zero (i.e. the initial direction coincides with the z-axis), the Fisher distribution reduces to

$$f_F(\theta, \phi) = \frac{\kappa}{2\pi \sinh \kappa} \sin \theta \; \cosh[\kappa \cos\theta] \quad (3.37b)$$

When κ tends towards 0, the Fisher distribution tends towards the isotropic one. When κ becomes very large, all the normals are aligned parallel to the polar direction. Often the prefactor will be denoted as $\mathcal{B}$

$$\mathcal{B} = \frac{\kappa}{2\pi \sinh \kappa} \quad (3.37c)$$

The angle ω (see Fig. 3.4) which is irrelevant for disks is usually assumed to be uniformly distributed over the interval $[0, 2\pi]$.

All the following results for homogeneous networks are independent of the individual fracture shapes and sizes, provided that they are not correlated with their orientations.

The quantities, whose general expressions are listed in Table 3.2 can be expressed analytically for anisotropic networks which obey Fisher's distribution. These calculations are the reserve of the aficionados of Bessel functions (a vanishing species, unfortunately). The following correction factors are defined as the ratios between a given average quantity and its value in the I^{2}OUD reference case

$$\Phi = \frac{V_{ex}}{V_{ex,r}}, \quad \psi_\ell = \frac{\langle n_I \rangle}{\langle n_{I,r} \rangle}, \quad \psi_t = \frac{\Sigma_t}{\Sigma_{t,r}}, \quad \psi_c = \frac{\mathcal{C}}{\mathcal{C}_r}, \quad \psi_p = \frac{\Sigma_p}{\Sigma_{p,r}} \quad (3.38)$$

Note that it is useless to introduce a specific correction factor for $\langle c \rangle$ since by definition eqn 3.21 implies

$$\frac{\langle c \rangle}{\langle c \rangle_r} = \frac{\psi_c}{\psi_t} \quad (3.39)$$

The correction Φ to the excluded volume is expressed as

$$\Phi = \frac{4}{\pi} \langle \sin \beta_{12} \rangle = \frac{2}{\sinh^2 \kappa} \left[I_0(2\kappa) - \frac{1}{\kappa} I_1(2\kappa) \right] \quad (3.40)$$

where I_0 and I_1 are the modified Bessel functions; β_{12} is the angle between the normal vectors of two fractures.

When the vector $\boldsymbol{p}$ is supposed to coincide with the polar direction $\boldsymbol{p}_F$ of the Fisher PDF and therefore with the z-axis, one obtains

$$\psi_\ell = 2\left(1 + \frac{1 - \cosh\kappa}{\kappa\sinh\kappa}\right) \quad (3.41)$$

When the normal $\boldsymbol{\nu}_\Pi$ to the plane Π is supposed to coincide with the polar direction of the Fisher PDF and therefore with the z-axis, the result is

$$\psi_t = \frac{2}{\sinh\kappa} I_1(\kappa) = \psi_c, \quad \psi_p = 4\left[\frac{I_1(\kappa)}{\sinh\kappa}\right]^2 = \psi_c^2 \quad (3.42)$$

Therefore, (3.39) and (3.42) imply that the mean chord length is not affected by anisotropy.

This subsection can be ended by commenting on a relation which was demonstrated in a particular case by Berkowitz and Adler (1993)

$$\Sigma_p = \frac{\Sigma_t^2 \langle c \rangle^2}{\pi} \quad (3.43)$$

This relation is still valid in the present situation since the only requirement is that the traces should be isotropically distributed in the plane Π, which is indeed the case when the Fisher polar direction is perpendicular to it.

3.8.3 The percolation threshold of anisotropic networks of convex fractures

The mean number ρ' of intersections per fracture in a network with density ρ is deduced from Φ defined by (3.40) as

$$\rho' = \frac{1}{2} AP\,\Phi(\kappa)\,\rho \quad (3.44)$$

Mourzenko et al. (2009) observed that the percolation threshold in terms of ρ' varies only very slightly with κ. Numerical simulations with hexagonal and square fractures reveal a small decrease of ρ'_c when κ increases from zero, until a constant value is reached for $\kappa \geq 10$. This value is 2.24 ± 0.02 for hexagons and 2.20 ± 0.03 for squares. Furthermore, percolation occurs simultaneously along the directions parallel and perpendicular to the Fisher pole $\boldsymbol{p}_F$. Therefore, (3.11) can be used as a first approximation.

3.8.4 Stereological relations for anisotropic and heterogeneous networks of convex fractures

The spatial distribution of the fractures is characterized by its volumetric density $\rho(\boldsymbol{x})$ defined as the number of centers per unit volume. Of course, the simplest distribution corresponds to a constant density $\rho(\boldsymbol{x}) = \rho_o$. A typical spatial distribution of fractures is provided by the exponential law (3.2). Such distribution has been observed in excavated damaged zones by Bossart et al. (2002) and Thovert et al. (2011) where for symmetry reasons the excavated wall is perpendicular to the z-axis.

In this subsection, the spatial distribution of fractures is no longer uniform, but given by eqn 3.2. Generally speaking, the z-axis in eqn 3.2 is parallel to the polar direction $\boldsymbol{p}_F$ of the Fisher PDF which is given by eqn 3.37b. Most of the results are established for circular fractures with radius R. Then, the decay length ℓ_d in eqn 3.2 is made dimensionless by

$$\ell'_d = \ell_d/R \tag{3.45}$$

Consider two disks with normals $\boldsymbol{n}_1$ and $\boldsymbol{n}_2$, respectively. If the origin of the coordinate system is at the center of the first disk, then the position of the second one is $\boldsymbol{x}$. In this coordinate system, fracture density (3.2) can be expressed as

$$\rho_o \exp(-\boldsymbol{x} \cdot \boldsymbol{p}_F/\ell_d) \tag{3.46}$$

Moreover, the normals $\boldsymbol{n}_1$ and $\boldsymbol{n}_2$ conform with the Fisher PDF (3.37a) with the pole $\boldsymbol{p}_F$.

Mourzenko et al. (2011a) derived the excluded volume for these two disks. The number of fractures $\langle n_I \rangle$ which intersect a given line per unit length is expressed as

$$\langle n_I \rangle = \frac{1}{2L} \int_0^L A\rho(z)\mathrm{d}z\, \psi_l,$$
$$\text{with} \quad \psi_\ell = 4\ell'_d \frac{\kappa}{\sinh \kappa} \int_0^1 \mathrm{d}s I_1\left(\frac{s}{\ell'_d}\right) \cosh\left(\kappa\sqrt{1-s^2}\right) \tag{3.47}$$

The integrals in (3.47) cannot be calculated analytically and are evaluated numerically. The variations of ψ_l with ℓ'_d and κ are illustrated in Fig. 3.21a; the influence of ℓ'_d when it is larger than 1, is almost always negligible.

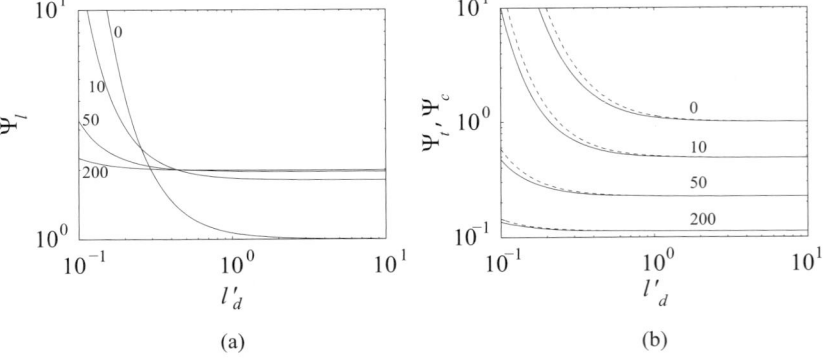

Fig. 3.21 The correction factors ψ_l (a), and ψ_t and ψ_c (b), for disks as functions of the dimensionless ratio ℓ'_d for various anisotropy coefficients κ. In (b), data are for: ψ_c (————), ψ_t (- - - -). The numbers denote the values of κ.

When the Fisher pole $\boldsymbol{p}_F$, the z-axis and the normal $\boldsymbol{\nu}_\Pi$ to the plane Π are all parallel and when the fractures are disks of radius R, Σ_t and ψ_t are expressed as

$$\Sigma_t = \frac{\rho_o \pi R}{2} \psi_t, \quad \psi_t = 2 \frac{\kappa}{E \sinh \kappa} I_1(E) \tag{3.48}$$

where

$$E = (\kappa^2 + \frac{1}{\ell_d'^2})^{1/2} \tag{3.49}$$

Let us now consider the total trace length $\mathcal{C}$ per unit surface for disks of radius R

$$\mathcal{C} = \rho_o R^2 \frac{\pi^2}{4} \psi_c, \quad \psi_c = \frac{2\kappa}{E \sinh \kappa} I_{01}(\kappa, E) \tag{3.50a}$$

with

$$I_{01}(\kappa, E) = \left[I_1\left(\frac{E+\kappa}{2}\right) I_0\left(\frac{E-\kappa}{2}\right) + I_0\left(\frac{E+\kappa}{2}\right) I_1\left(\frac{E-\kappa}{2}\right) \right] \tag{3.50b}$$

The variations of ψ_t and ψ_c with ℓ_d' and κ are illustrated in Fig. 3.21b with the same conclusion as for ψ_l, namely the influence of ℓ_d' is negligible when it is larger than 1.

The average trace length $\langle c \rangle$ is equal to the total trace length (3.50) divided by the number of traces (3.48)

$$\langle c \rangle = \frac{\mathcal{C}}{\Sigma_t} = \frac{\pi R}{2 I_1(\kappa)} I_{01}(\kappa, E) \tag{3.51}$$

Σ_p is straightforwardly estimated from (3.43)

$$\Sigma_p = \frac{\rho_o^2 R^4 \pi^3}{16} \psi_p, \quad \psi_p = \frac{4\kappa^2}{E^2 \sinh^2 \kappa} I_{01}^2(\kappa, E) = \psi_c^2 \tag{3.52}$$

3.8.5 Networks of heterogeneous fractures

In this subsection, a heterogeneous fracture is defined as a fracture whose aperture is variable. In order to schematize these variations, Hamzehpour et al. (2009) assumed that the aperture could be equal to two values: b and 0. Therefore, the statistical generation of such an aperture field in each fracture is equivalent to the generation of a phase function $Z(\boldsymbol{x})$ which is equal to 1 when the aperture is equal to b; otherwise $Z(\boldsymbol{x})$ is equal to zero. $Z(\boldsymbol{x})$ can correspond to the phase function $1 - Z_c(\boldsymbol{x})$ of the open regions in a fracture (see eqn 2.22 and Fig. 2.5). This phase function is statistically characterized by a probability S_0 which is the fractional open area (2.22c), and by a correlation function $\mathcal{R}_Z(u)$

$$S_0 = \overline{Z(\boldsymbol{x})}, \quad \mathcal{R}_Z(u) = \frac{\overline{[Z(\boldsymbol{x}) - S_0][Z(\boldsymbol{x} + \boldsymbol{u}) - S_0]}}{(S_0 - S_0^2)} \tag{3.53a}$$

where u is the norm of the translation vector $\boldsymbol{u}$. The overbar denotes the spatial average. $Z(\boldsymbol{x})$ is derived by thresholding a Gaussian field $Y(\boldsymbol{x})$ correlated by

$$\mathcal{R}_Y(u) = \exp\left[-\frac{u^2}{\ell_c^2}\right] \tag{3.53b}$$

where ℓ_c is the correlation length (cf. eqn 2.8). This generation process for Z corresponds exactly to the generation of a 2D porous medium as detailed by Adler (1992) and the generation of the field Y corresponds to that of a random surface as described in Section 2.3.1.

A heterogeneous fracture and a fracture network made of heterogeneous fractures are shown in Fig. 2.5c and Fig. 3.9c, respectively. Note that in this section, the fractures are IOUD as in the reference case; more precisely, they are identical in size and shape, but the distribution of open regions varies from fracture to fracture.

The qualitative influence of the probability S_0 and of the fracture density ρ can be schematized. Two fractures may intersect, but in order for the fluid to flow from one to another, the open zones of each fracture need to overlap at least partially along the intersections; otherwise, the two fractures are not connected when flow is considered. Of course, this is not sufficient in that these two open regions should also belong to the percolating cluster.

Therefore, when the network of fractures of constant aperture b percolates for a given fracture density ρ, one can determine the minimal value S_{0c} for which the network of heterogeneous fractures percolates; S_{0c} is expected to be a decreasing function of ρ or equivalently of ρ'.

S_{0c} should also depend on the second physical parameter ℓ_c/R. The study of this dependence necessitates systematic numerical calculations for various values of ℓ_c/R which can be summarized by the following equation

$$S_{0c} = S_{0c}(\rho', \ell_c/R) \tag{3.54}$$

The limit $\ell_c/R = \infty$ is known since the network is composed of conducting fractures with a spatial density $S_0\rho'$; therefore, the former result (3.10) of Huseby et al. (1997) can be used to derive the expression

$$S_{0c} = 2.3\rho'^{-1} \quad \text{for} \quad \ell_c/R = \infty \tag{3.55}$$

In addition to this known limit, two other values were selected for ℓ_c/R, namely 0 and 1. The major data and the intermediate steps are identical. A tentative semi-empirical law with no theoretical background can be proposed in order to synthesize the data

$$S_{0c} = \beta \rho'^{\alpha} \tag{3.56}$$

where β and α are parameters which depend on ℓ_c/R. When (3.55) is used, α and β can be expressed as

$$\alpha = -1 + 0.68 e^{-0.81(\ell_c/R)}, \tag{3.57a}$$

$$\beta = 2.3 - 1.65 e^{-0.82(\ell_c/R)}. \tag{3.57b}$$

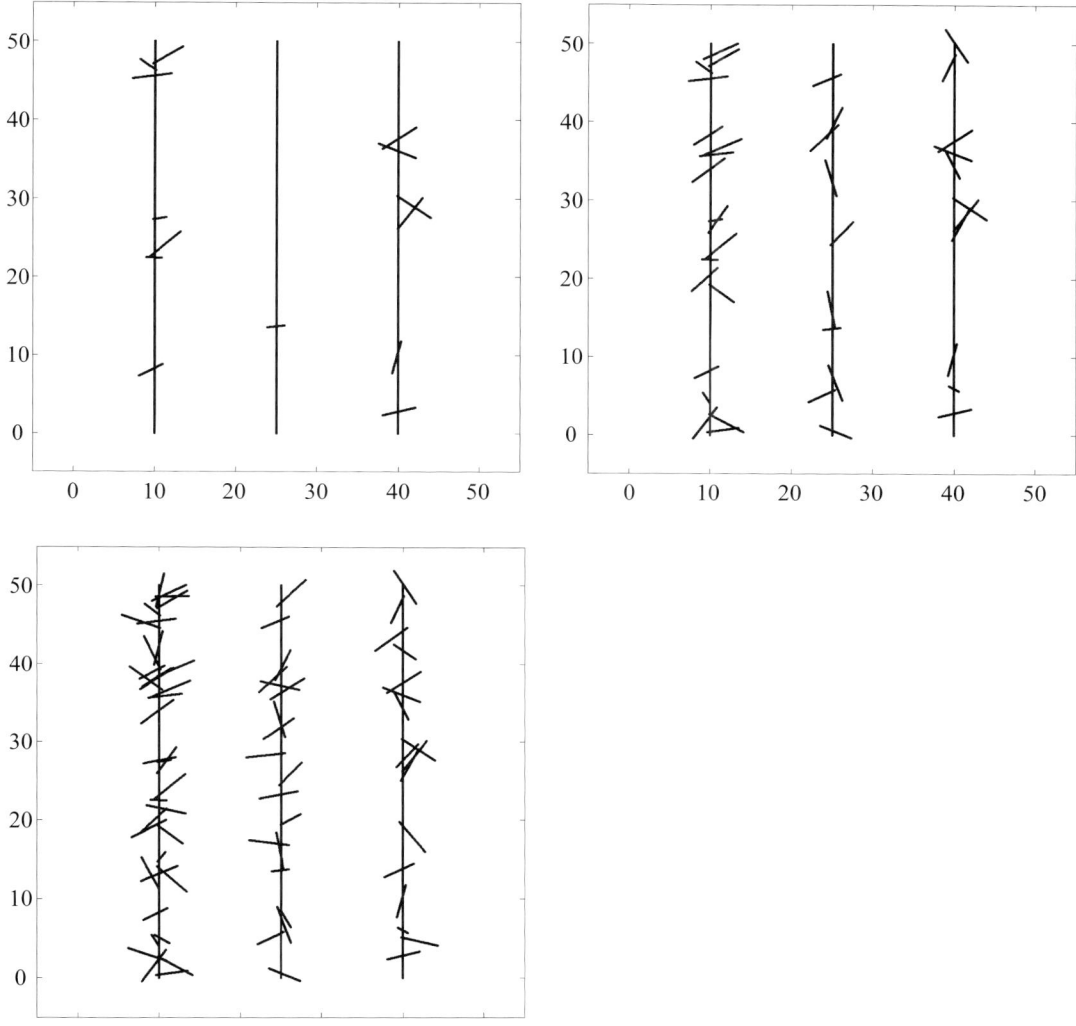

Fig. 3.22 Three outcrops for Exercise 3.7. Only the fracture traces (short lines) which intersect the scanlines (thick long lines of length 50 m) are reported. Dimensions are in meters.

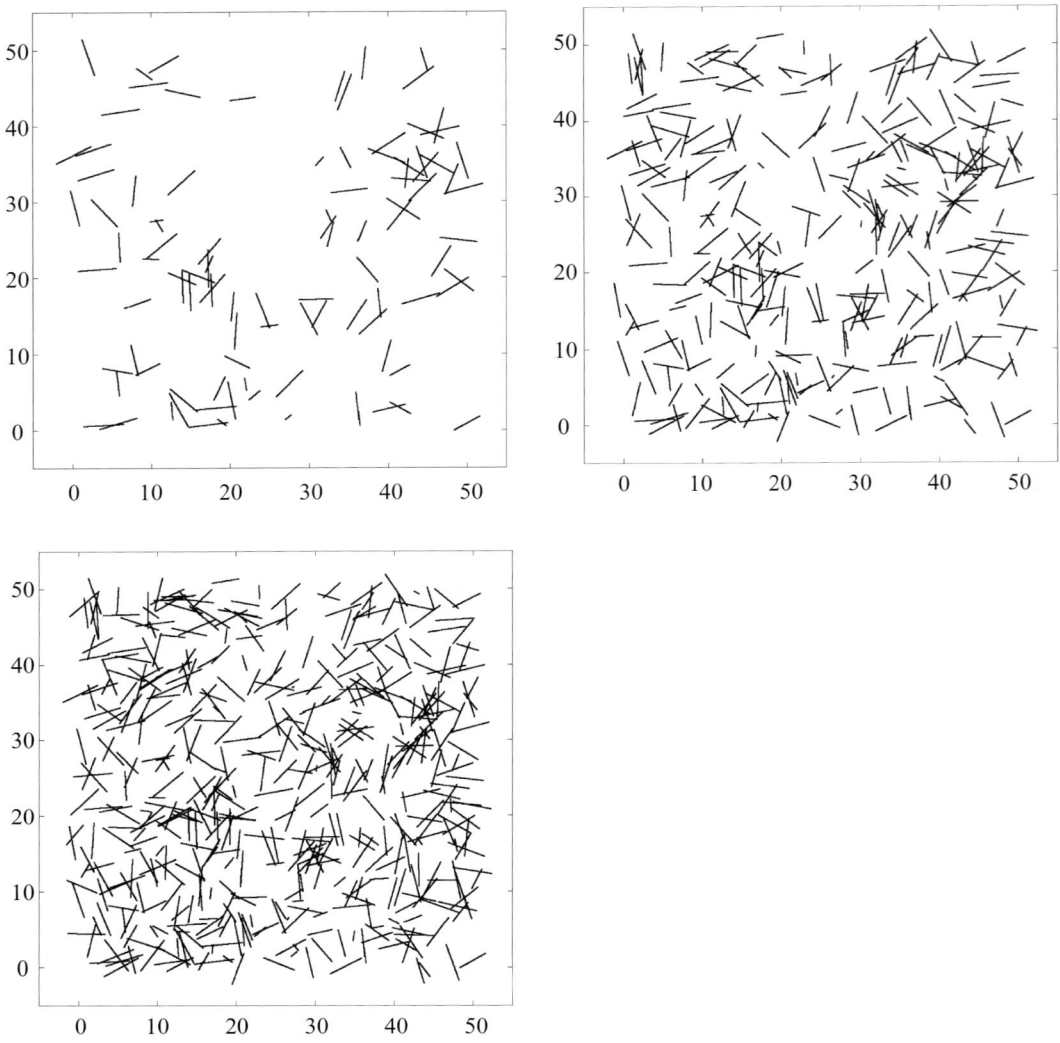

Fig. 3.23 The three complete outcrops which were only partially provided in Fig. 3.22. Dimensions are in meters.

Exercises

(3.1) (i) Read the general calculation of the excluded volume pp. 189–193 by Adler and Thovert (1999) and redo it.

(ii) Derive by yourself the excluded volume of two disks.

(3.2) (i) Determine $V_{ex,r}$ for two equal disks of radius R by a direct application of (3.5b).

(ii) Calculate the dimensional critical fracture density when $R_d = 1$ m, 0.1 m.

(iii) Comment.

(iv) Same three questions for two equal squares of side $a = 2R_d$.

(3.3) Calculate $\rho'_{c,r}$ for the six shapes used by Huseby et al. (1997), namely equilateral triangles, squares, hexagons, octagons, 20-gons and rectangles with an aspect ratio equal to 2. Explain why they found $\rho'_{c,r} \approx const$.

(3.4) Determine analytically the excluded surface of two segments of lengths l_1 and l_2 isotropically oriented and uniformly distributed in the plane. Hint. First, consider a fixed relative orientation; second, average over the orientations.

(3.5) Consider disks of radius $R = 1$ m and squares of sides $a = 2$ m. Determine ρ so that $\rho_{b,r} = 1$ m^{-3}.

(3.6) Start from (3.17) and derive (3.14a) for an isotropic distribution of fractures.

(3.7) Three outcrops are provided in Fig. 3.22. Dimensions are in meters. In each outcrop, three scanlines of 50 m in length are drawn as thick lines. The short segments correspond to the traces which intersect the scanlines. This exercise is purposely incomplete so that the reader can make hypotheses in order to answer the questions. In a way, it mimics real situations where one never has complete knowledge of a real fracture field.

(i) Analyze the data measured along the three scanlines in the three different outcrops which are displayed in Fig. 3.22. Determine ρ, the percolating character or not of the networks, $\rho_{b,r}$.

(ii) Analyze the corresponding complete trace maps displayed in Fig. 3.23. Derive $\Sigma_{t,r}$ and ρ.

(iii) Discuss and comment.

(3.8) Calculate ζ_3 for the trace maps of Fig. 3.23.

(3.9) The total probability for all the fracture sizes should be equal to 1.

(i) Derive the normalization condition for a power-law distribution and relate a to the extremal radii. In other words, derive eqn 3.27.

(ii) Determine v_{ex} and $<R^3>$ for circular disks whose radii are distributed according to this power law.

(iii) Calculate ρ_c for $a = 2$, $R_m = 0.1$ m, and $R_M = 1$ m.

4 Transport in a single fracture

4.1 Introduction 66
4.2 Flow of a Newtonian fluid 66
4.3 Diffusion of a passive solute 74
4.4 Dispersion of a passive solute 78
4.5 Extensions 81
Exercises 85

4.1 Introduction

In this chapter, we revert to the scale symbolised by b, the local aperture which was defined in Chapter 2 by eqn 2.21. In other words, one can again clearly see the two surfaces which limit the void space.

The main general objective of this chapter is to show how the local transmissivity of a fracture can be derived from the Stokes equations when its geometrical characteristics are known, and how this approach can be extended to other phenomena such as conduction and dispersion.

There was also an important objective when these calculations were made in the mid 1990s, namely to show that the lubrication approximation which was widely used in the literature was totally inadequate. Actually, as will be shown in Section 4.2, this approximation demands conditions which cannot be fulfilled in real fractures.

This chapter is organized as follows. Section 4.2 details the analysis of flow and then, conductivity and dispersion of a passive solute are briefly presented in Sections 4.3 and 4.4. As was the case previously, this chapter ends with extensions which are mainly devoted to self-affine fractures.

4.2 Flow of a Newtonian fluid

4.2.1 Rigorous approach

The following analysis can be readily applied to a flow experiment performed in a laboratory on a single fracture of length L and width W. The fracture could be artificially generated by a Brazilian test performed on a solid cylinder; the cylinder is wrapped in an impermeable material and two different pressures are applied to the two ends of the fracture (cf. Fig. 1.4). A more specific view is provided by Fig. 4.1.

The low Reynolds number flow of an incompressible Newtonian fluid is governed by the Stokes equations and boundary conditions which are recalled here

$$-\nabla p + \mu \nabla^2 \boldsymbol{v} = 0 \qquad (4.1a)$$

$$\nabla \cdot \boldsymbol{v} = 0 \qquad (4.1b)$$

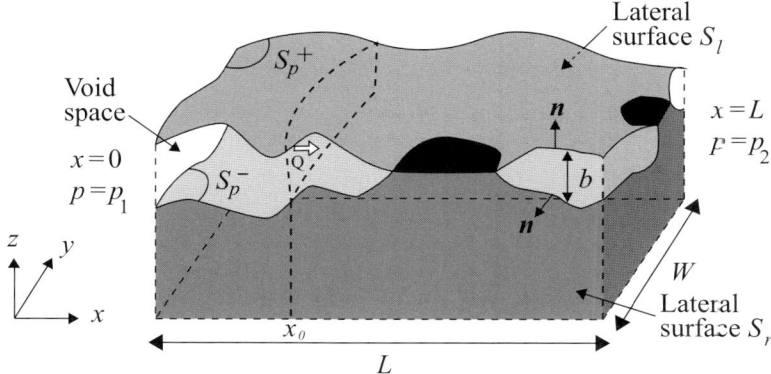

Fig. 4.1 Experimental and numerical determination of the transmissivity of a fracture of length L and width W. The solid surfaces $S_p^\pm$ limit the void space τ where flow takes place.

where p and $\boldsymbol{v}$ denote pressure and velocity, respectively; μ is fluid viscosity.

This system should be supplemented by overall boundary conditions at the external fracture boundary, as illustrated in Fig. 4.1. Assuming that the solid matrix is impermeable, then fluid velocity should vanish on the two solid surfaces $S_p^\pm$ which limit the void space. In addition, the fracture is wrapped in an impermeable container or material and the two lateral surfaces S_l and S_r are also surfaces where fluid velocity should vanish. Finally, the driving force is the pressure difference exerted between the upstream and the downstream surfaces at $x = 0$ and L; one can imagine that the two ends of the fracture are put in two large vessels where pressure is more or less constant. These boundary conditions can be formalized as

$$\boldsymbol{v} = 0 \quad \text{on} \quad S_p^+, \ S_p^-, \ S_l \quad \text{and} \quad S_r \qquad (4.2a)$$

and

$$p = p_1 \quad \text{at} \quad x = 0; \ p = p_2 \quad \text{at} \quad x = L \qquad (4.2b)$$

These equations can be commented as follows. First, one should not forget that they are only valid for low Reynolds numbers, i.e. when the inertial forces are small compared to the viscous forces. Within this limit, the equations are linear and the flow rate is expected to be proportional to the pressure difference as discussed in Section 1.3.

Second, the previous problem can be posed in the rigorous framework of homogenization theory. The sample is the unit cell of an infinite two-dimensional spatially periodic medium. Indeed, these are the conditions where the numerical calculations are actually done, but the previous presentation, which mimics the experimental procedure, is much simpler to explain and provides all the necessary ingredients.

Third, as for the complex surfaces introduced in Chapter 2, there is no analytical solution and one has to use numerical techniques; those developed for porous media and detailed by Adler (1992) are perfectly suited for this purpose.

When these equations are solved, one can calculate the overall flow rate Q along the x-axis by integrating the local velocity field over a cross section $x = x_0 =$ constant normal to the x-axis (see Fig. 4.1)

$$Q = \int\int_{x_0} u \, dy dz \qquad (4.3)$$

where u is the x-component of the fluid velocity.

For a porous medium, the easily measurable quantity is the flow rate per unit surface as shown in Fig. 1.4. However, for a fracture, this quantity is the flow rate J per unit fracture width; indeed, the width W of the fracture is easy to measure and unambiguous in a laboratory experiment; note that the contact zones are also taken into account. J enables one to define a *fracture transmissivity* σ

$$J = \frac{Q}{W} = -\frac{\sigma}{\mu}\frac{p_2 - p_1}{L} \qquad (4.4a)$$

or in vectorial terms

$$\boldsymbol{J} = -\frac{\boldsymbol{\sigma}}{\mu} \cdot \overline{\nabla p} \qquad (4.4b)$$

where $\boldsymbol{\sigma}$ is a 2×2 tensor. $\overline{\nabla p}$ is a macroscopic pressure gradient defined on a large scale with respect to a typical fracture aperture b; the overbar is often forgotten, in order to simplify notations. Moreover, since the fracture is globally parallel to the xy-plane, the gradient operator ∇ is two-dimensional. Equation 4.4b is a two-dimensional counterpart of Darcy's law (1.5).

$\boldsymbol{\sigma}$ can be commented as follows. $\boldsymbol{\sigma}$ shares many of the properties of the standard permeability $\boldsymbol{K}$. $\boldsymbol{\sigma}$ only depends on the geometry of the fracture, but it is homogeneous to the cube of a length. Only isotropic fractures are considered in this book and the fracture transmissivity $\boldsymbol{\sigma}$ is reduced to a spherical tensor $\sigma \boldsymbol{I}$ where $\boldsymbol{I}$ is the unit 2×2 tensor. Often the subscript S is added in order to indicate that σ is calculated by means of the Stokes equation. Therefore,

$$\boldsymbol{J} = -\frac{\sigma_S}{\mu}\nabla p \qquad (4.4c)$$

There is a simple solution to the Stokes equations, which is the Poiseuille flow, i.e. the flow between two parallel planes separated by a constant distance b_m. It can be said that this is the only solution which can be easily worked out, a fact which is not entirely true, but far from wrong!

This elementary situation is demonstrated in Fig. 4.2. Assume that flow occurs between these walls in straight lines which are parallel to the x-axis and that the z-axis is perpendicular to the walls. The Stokes equations (4.1) reduce to

$$\mu\frac{\partial^2 u}{\partial z^2} = \frac{dp}{dx}, \quad \frac{\partial u}{\partial x} = 0 \qquad (4.5)$$

where u is the velocity component along x which must comply with the adherence condition

$$u = 0 \quad \text{at} \quad z = 0 \quad \text{and} \quad b_m \qquad (4.6)$$

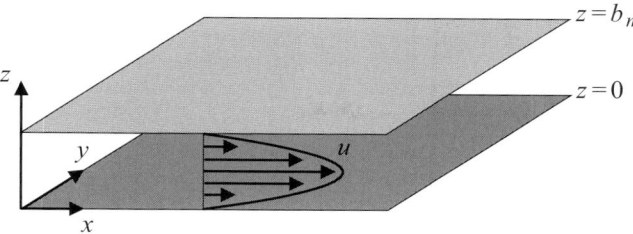

Fig. 4.2 The Poiseuille flow between two parallel planes which are separated by a constant distance b_m. The parabolic flow profile is indicated by the thick solid line.

Since u does not depend on x, dp/dx can be considered as a constant. The dynamic equation can be integrated as

$$u = -\frac{1}{2\mu}\frac{dp}{dx}z(b_m - z) \tag{4.7}$$

a parabolic distribution known as the *Poiseuille flow*. The total flow rate per unit width of plate J is obtained by integrating (4.7)

$$J = \frac{1}{b_m}\int_0^{b_m} u\,dz = -\frac{b_m^3}{12\mu}\frac{dp}{dx} \tag{4.8}$$

Therefore, the fracture transmissivity σ_P for a Poiseuille flow is

$$\sigma_P = \frac{b_m^3}{12} \tag{4.9}$$

We can conclude this subsection with comments on the appropriate terminology for a fracture flow coefficient. There is some confusion regarding terms such as "hydraulic conductivity", "permeability" and "transmissivity" which are often used with different meanings. Barton *et al.* (1985) and Zimmerman *et al.* (1991) use "hydraulic conductivity" to denote the quantity σ in (4.4). "Permeability" is used when the fracture is viewed as a layer of thickness b of permeable material (Brown, 1989; Zimmerman and Bodvarsson, 1996 among others), and corresponds to σ/b. It can correspond to the permeability K_f in eqn 6.5c of an actual filling material, or to $b_m^2/12$ for an empty channel (4.9). However, Witherspoon *et al.* (1980) refer to this quantity as "hydraulic conductivity". Fracture permeability is also often used with a totally different meaning, which corresponds to the contribution of a fracture to the macroscopic permeability of the fractured rock (Barton *et al.*, 1985), especially in commercial computer codes based on a double permeability model. Finally, "transmissivity" always corresponds to σ in eqn 4.4 (Zimmerman and Bodvarsson, 1996; Méheust and Schmittbuhl, 2001 among others). This concept originally quantifies how much fluid can be transferred by a geological formation such as a confined aquifer or a production layer, and can be defined and measured without any knowledge or hypothesis of its thickness and contents. Its use is advocated for fractures by the US National Research Council Committee on Fracture Characterization and Fluid Flow (1996) who also wisely emphasizes "the importance of reporting the test results with sufficient detail on the method of analysis so that readers are not mislead".

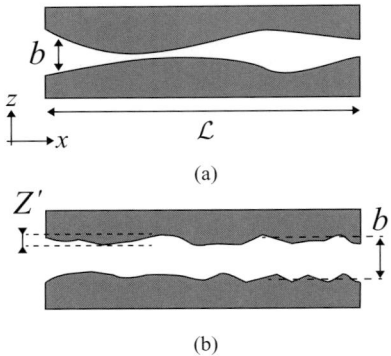

Fig. 4.3 The two situations which can be addressed by means of analytical approximations.

4.2.2 Analytical approximations

As a first approximation, it is natural to view a fracture as a plane channel of constant aperture. This viewpoint has the advantage of providing explicit solutions which are easy to derive. However, such a simple representation is obviously not adequate for the structures presented in Section 2.2.

One can proceed further with analytical calculations in two different cases which are both characterized by the existence of a small parameter ϵ. In the first case, the variations along the x-axis are very slow when compared to the variations along the two perpendicular directions, as depicted in Fig. 4.3a; there might be a significant decrease in the aperture b of the channel, but this occurs over a characteristic distance $\mathcal{L}$ so that

$$\mathcal{L} \gg b, \quad \text{i.e.} \quad \epsilon = \frac{b}{\mathcal{L}} \ll 1 \qquad (4.10)$$

This case can be seen as a physical situation with no statistical homogeneity; it can also correspond to a sinusoidal channel with non-negligible aperture variations which occur over very long distances.

The second type of approximation depicted in Fig. 4.3b is characterized by the fact that the amplitudes of the wall oscillations, say Z', are very small when compared to the mean channel aperture b

$$b \gg Z', \quad \text{i.e.} \quad \epsilon = \frac{Z'}{b} \ll 1 \qquad (4.11)$$

In terms of waves, eqn 4.10 corresponds to very long wavelengths with possible large amplitudes, while eqn 4.11 corresponds to small amplitudes with possible small wavelengths.

These conditions need to be compared to the experimental findings summarized by eqn 2.12. σ_h can be roughly compared to the amplitudes of the wall oscillations and $\mathcal{L}$ to the correlation length ℓ_c. Therefore, at this point, it is clear that real fractures do not comply with lubrication conditions at all. Despite this fact, all the early contributions on flow through fractures used the lubrication approximation and the first paper which went beyond this approximation in three dimensions is that of Mourzenko et al. (1995).

Examples of analytical flow calculations in the situation (4.11) are provided in Section 4.5.1.

The *lubrication approximation* which is valid when eqn 4.10 is satisfied, is used below to obtain the Reynolds equation. The equation which governs the pressure p, is derived from the continuity equation. Consider a portion of fracture which is limited by the vertical surface $\mathcal{S}$ whose intersection with the plane $z = 0$ is the curve $\mathcal{C}$ with the normal vector $\boldsymbol{n}$ (cf. Figs 4.4a and b). Integration of the continuity equation (4.1b) over the volume limited by $\mathcal{S}$, and use of the divergence theorem and of the no slip condition (4.2a) result in

$$\int\!\!\int_{\mathcal{S}} \boldsymbol{v} \cdot \boldsymbol{n} \, \mathrm{d}s = \int_{\mathcal{C}} \boldsymbol{J} \cdot \boldsymbol{n} \, \mathrm{d}\ell = 0 \qquad (4.12)$$

where $\mathrm{d}\ell$ is the line differential element.

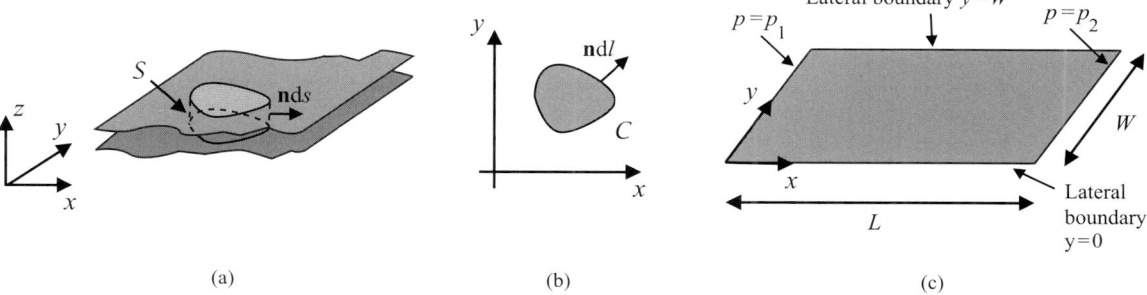

Fig. 4.4 Definitions. Mass conservation (a) and (b). Boundary conditions for a fracture in the Reynolds approximation (c).

Since this relation should be valid for any surface $\mathcal{S}$ (or curve $\mathcal{C}$), one obtains

$$\nabla \cdot \boldsymbol{J} = 0 \qquad (4.13)$$

It should be noted that the operator ∇ is two-dimensional in the rest of this section, starting from the last equation.

The flux $\boldsymbol{J}$ is approximated locally by the Poiseuille formula (4.8)

$$\boldsymbol{J} = -\frac{b^3}{12\mu}\nabla p \qquad (4.14)$$

By combining eqns 4.13 and 4.14, one obtains

$$\nabla \cdot (b^3 \nabla p) = 0 \qquad (4.15a)$$

This equation should be supplemented with the following boundary conditions illustrated in Fig. 4.4c

$$p = p_1 \quad \text{at} \quad x = 0; \quad p = p_2 \quad \text{at} \quad x = L \qquad (4.15b)$$

and a no-flux condition on the two lateral sides

$$\frac{\partial p}{\partial y} = 0 \quad \text{at} \quad y = 0, W \qquad (4.15c)$$

This system is the so-called Reynolds approximation. It is much simpler to solve than the three-dimensional Stokes equation with four unknowns at each point since it is two-dimensional with the pressure as the only unknown.

A second fracture transmissivity σ_R can be defined by using this flow determination

$$J = -\frac{1}{W}\int_0^W \frac{b^3}{12\mu}\frac{\partial p}{\partial x}(x=0, y)\,\mathrm{d}y = -\frac{\sigma_R}{\mu}\overline{\nabla p} \qquad (4.16)$$

This subsection can be summarized by stating that two overall fracture transmissivities were defined according to the calculation method, namely the Stokes and the Reynolds permeabilities denoted by σ_S and

σ_R which are solutions to systems (4.1) and (4.15), respectively. System (4.15) is much easier to solve than (4.1), but is quite approximate.

One can also define equivalent mean apertures by using the cubic law, i.e. by inverting eqn 4.9

$$b_S = (12\sigma_S)^{1/3}, \quad b_R = (12\sigma_R)^{1/3} \qquad (4.17)$$

4.2.3 Estimations of the fracture transmissivity σ

Suppose that for some reason you are abandoned on a desert island. That reason is beyond the scope of this book! Suppose also that for some (probably different) reason you need to estimate the transmissivity of a fracture. The best thing to do if you do not have this book with you (we are now beginning to understand why you are on this island) is to rederive the Poiseuille equation (4.9)!

If one reflects a little more, one remembers (cf. Section 2.2.3) that the geometry of a fracture can be characterized by its average aperture b_m, its roughness σ_h, the nature of its surface (Gaussian, self-affine, ℓ_c, H and possibly other characteristics) and the intercorrelation coefficient θ_I. Since σ_S only depends on the geometry, one can write

$$\sigma'_S = \frac{\sigma_S}{\sigma_{av}} = f\left(\frac{b_m}{\sigma_h}, \frac{\ell_c}{\sigma_h}, \theta_I, \text{fracture nature}\right) \qquad (4.18a)$$

where

$$\sigma_{av} = \frac{\langle b \rangle^3}{12} \qquad (4.18b)$$

This choice of the unit σ_{av} for transmissivity is somewhat arbitrary and another unit such as $b_m^3/12$ could have been chosen. The same unit is chosen for σ'_R.

But you do have this book with you! So far, fractures could be either Gaussian or self-affine. From now on, fractures have a Gaussian correlation provided by (2.8). Self-affine fractures are addressed in Section 4.5. Fig. 4.5 displays some master curves that could be useful. Let us comment on them. In principle, σ'_S should be determined for an infinite fracture, but as this is not feasible, it is calculated as the statistical average over a sufficiently large number of fractures; five independent realizations were considered sufficient. A question that one may ask concerns the potential influence of the correlation coefficient. It is clear in (a) that the true transmissivity σ'_S increases slightly with θ_I for $\ell_c/\sigma_h = 1$; actually, the influence of this parameter on transmissivity can be neglected in a first approximation for low values of ℓ_c/σ_h. This is no longer the case for $\ell_c/\sigma_h = 4$ though the effect is great only for small values of b_m/σ_h.

It is remarkable to observe in the same figure that the Reynolds approximation depends on θ_I, and not on ℓ_c/σ_h. This last feature can be explained by the fact that an increase of ℓ_c/σ_h corresponds to an increase in the discretization of the fracture, but not in a change in the fracture's physical parameters. A similar phenomenon has already been observed for correlated porous media.

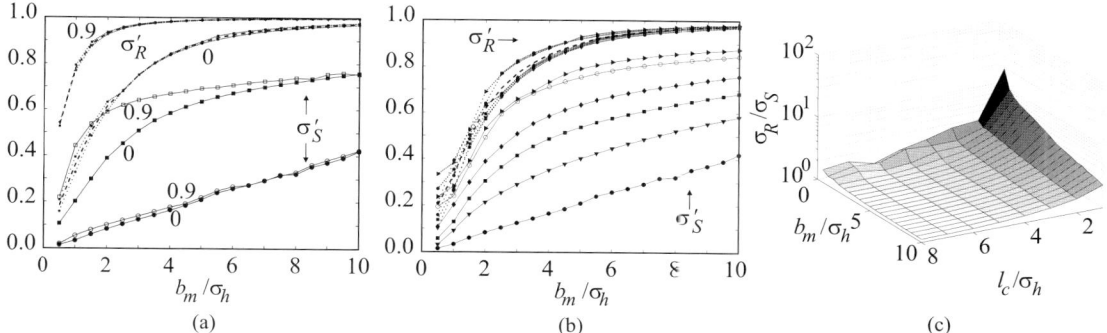

Fig. 4.5 Dimensionless transmissivities σ'_S (solid lines) and σ'_R (broken and dotted lines) of Gaussian fractures. The values are transmissivity averages over five independent realizations along the x- and y-axes; the size of the unit cell is $160a$ where a is the size of the elementary cube; $\sigma_h = 5a$. (a) Influence of the intercorrelation coefficient θ_I. Data are for: $\ell_c/\sigma_h = 1$ ($\theta_I = 0$, •; $\theta_I = 0.9$, ○), $\ell_c/\sigma_h = 4$ ($\theta_I = 0$ ■; $\theta_I = 0.9$ □). (b) σ'_S and σ'_R as functions of b_m/σ_h; $\theta_I = 0$; data are for: $\ell_c/\sigma_h=1$ •, 2 ▼, 3 ■, 4 ♦, 6 *, 8 ▶. (c) The ratio σ_R/σ_S as a function of b_m/σ_h and ℓ_c/σ_h for $\theta_I = 0$.

Moreover, the Reynolds approximation is not that good, since it sometimes overestimates the true transmissivity σ'_S calculated with the Stokes equation by almost an order of magnitude. We shall come back to this point in the next subsection.

Systematic results are provided in Fig. 4.5b for mutually uncorrelated fracture surfaces. The Reynolds transmissivity is also given. σ'_S increases with b_m/σ_h and with ℓ_c/σ_h. These two variations can be interpreted in terms of the roughness σ_h; when σ_h diminishes, the fluid flows more easily through the fracture and transmissivity increases, as observed.

It is seen that ℓ_c/σ_h has a significant influence on σ'_S. Another interesting feature is that σ'_S is seen to tend towards σ'_R when ℓ_c/σ_h increases.

4.2.4 Comparison between Reynolds and Stokes transmissivities

In view of its importance, it is better to devote a complete subsection to this comparison. We should recall that in Chapter 2 experiments show that typical values are (Gentier, 1986; Brown et al., 1986)

$$0 < \frac{b_m}{\sigma_h} < 2.6, \quad 1 < \frac{\ell_c}{\sigma_h} < 7 \qquad (4.19)$$

The numerical results discussed in the previous subsection can be displayed in a more dramatic way in Fig. 4.5c. The foreground of the graph corresponds to large values for b_m/σ_h and ℓ_c/σ_h for which the lubrication approximation is valid (cf. eqns 4.10 and 4.11); therefore, the ratio σ_R/σ_S tends towards 1 as it should do.

However, the background of the graph corresponds to the range $b_m/\sigma_h < 2.6$, i.e. to real fractures. The discrepancy which is observed between the real Stokes value and the Reynolds approximation, may increase to 20 for $\ell_c/\sigma_h=1$ and $b_m/\sigma_h=0.5$.

Exercise 4.1 provides a direct application of the results provided in this section.

4.3 Diffusion of a passive solute

Basically, the same developments as before can be produced for phenomena governed by the Laplace equation, namely for heat or electrical conduction and diffusion. This section is largely based on Volik *et al.* (1997).

4.3.1 A rigorous approach

For convenience sake, the local governing equations are recalled below. The local flux $\boldsymbol{j}$ is given by Fick's law

$$\boldsymbol{j} = -D\nabla c \qquad (4.20)$$

where D is the molecular diffusion coefficient and c is the volumetric solute concentration. In a context of heat (resp. electric) conduction, D is diffusivity (resp. electrical conductivity) and c is temperature (resp. potential). Mass conservation implies that

$$\frac{\partial c}{\partial t} + \nabla \cdot \boldsymbol{j} = \frac{\partial c}{\partial t} - \nabla \cdot (D\nabla c) = 0 \qquad (4.21)$$

Only steady situations with constant D are considered in this section. Thus, (4.21) reduces to the Laplace equation

$$\nabla^2 c = 0 \qquad (4.22)$$

Finally, the solid is supposed to be impervious to the diffusive species which implies

$$\boldsymbol{n} \cdot \nabla c = 0 \quad \text{on} \quad S_p^{\pm} \qquad (4.23)$$

where $\boldsymbol{n}$ is the normal vector to the solid surface $S_p^{\pm}$ (cf. Fig. 4.1). This condition obviously applies to solute diffusion, and it is generally well verified for electrical conduction. It is less realistic for heat conduction since it is not generally true that the surrounding material can be considered as an insulating one.

As in Section 4.2.1, this system should be supplemented by overall boundary conditions at the external fracture boundary. The developments for flow also apply here. The notations are illustrated in Fig. 4.1. The fracture is wrapped in an impermeable container or material and the two lateral surfaces S_l and S_r are also surfaces where flux should vanish. The driving force is the concentration difference which exists between the upstream and downstream surfaces at $x = 0$ and L; one can imagine that the two ends of the fracture are put in two large vessels where the solute concentration is roughly constant. These boundary conditions can be formalized as

$$\boldsymbol{j} \cdot \boldsymbol{n} = 0 \quad \text{on} \quad S_p^+, \ S_p^-, \ S_l \quad \text{and} \quad S_r \qquad (4.24\text{a})$$

and
$$c = c_1 \quad \text{at} \quad x = 0; \quad c = c_2 \quad \text{at} \quad x = L \tag{4.24b}$$

These equations can be commented on in the same way as in Section 4.2.1. The equations are linear too which means that each quantity is proportional to the concentration difference. The homogenization theory can be invoked to make everything more complicated. Of course, there is no analytical solution for the complex surfaces introduced in Chapter 2.

The total solute flux Q_c along the x-axis can be derived by integrating the local flux over any cross section $x = x_0 =$ constant

$$Q_c = \int\int_{x_0} j_x \, dy dz \tag{4.25}$$

Again it is the flux per unit width J_c which makes sense for a fracture and the following relation can be used to define the fracture conductivity Λ

$$J_c = \frac{Q_c}{W} = -\Lambda \frac{\partial c}{\partial x} \quad \text{with} \quad \frac{\partial c}{\partial x} = \frac{c_2 - c_1}{L} \tag{4.26a}$$

or in tensorial terms

$$\boldsymbol{J}_c = -\boldsymbol{\Lambda} \cdot \overline{\nabla c} \tag{4.26b}$$

As the pressure gradient $\overline{\nabla p}$ defined by (4.4b), $\overline{\nabla c}$ is a macroscopic concentration gradient defined on a scale large with respect to the typical fracture aperture b. As for an isotropic fracture, the 2×2 conductivity tensor $\boldsymbol{\Lambda}$ becomes a spherical tensor $\Lambda \boldsymbol{I}$ where $\boldsymbol{I}$ is the unit 2×2 tensor. One often adds the subscript L in order to indicate that Λ_L is deduced from the solution to the Laplace equation. Λ_L/D is homogeneous to a length and it only depends on the geometry of the fracture.

An elementary solution can be derived when the fracture is a channel limited by two parallel walls as illustrated by Fig. 4.2. The local flux $\boldsymbol{j}$ is parallel to the x-axis. Therefore, (4.22) is reduced to

$$\frac{\partial^2 c}{\partial x^2} = 0 \quad \text{with} \quad \frac{\partial c}{\partial z} = 0 \quad \text{at} \quad z = 0 \quad \text{and} \quad b_m \tag{4.27a}$$

The solution to these equations is

$$c(x, y, z) = \overline{\frac{\partial c}{\partial x}} x \tag{4.27b}$$

The corresponding macroscopic diffusion coefficient Λ_P for the Poiseuille configuration is

$$\Lambda_P = D b_m \quad \text{or} \quad \frac{\Lambda_P}{D} = b_m \tag{4.27c}$$

4.3.2 The Reynolds approximation

As in eqn 4.14, we assume that the fracture is defined by its aperture $b(x, y)$. Because of eqn 4.27c, the local flux per unit width is approximated by (subscript R for Reynolds)

$$\boldsymbol{J}_R = -D b(x, y) \nabla c \tag{4.28}$$

If the molecular diffusion coefficient is assumed to be constant (cf. 4.15a), mass conservation in the fracture implies

$$\nabla \cdot \mathbf{J}_R = -D\nabla \cdot [b(x,y)\nabla c] = 0 \qquad (4.29a)$$

Therefore, the 3D diffusion equation (4.22) and the boundary conditions (4.24) can be approximately reformulated as (4.29a) to be solved with the boundary conditions

$$c = c_1 \text{ at } x = 0; \quad c = c_2 \text{ at } x = L \qquad (4.29b)$$

and a no-flux condition on the two lateral sides (cf. Fig. 4.4)

$$\frac{\partial c}{\partial y} = 0 \quad \text{at} \quad y = 0, W \qquad (4.29c)$$

Again, this is called the *Reynolds approximation*. It is much simpler since the 3D Laplace equation (4.22) is replaced by a 2D Laplace equation (4.29).

A second fracture conductivity Λ_R can be defined by using this concentration determination

$$J_{cR} = -\frac{1}{W}\int_0^W Db\frac{\partial c}{\partial x}(x=0,y)dy = -\Lambda_R \overline{\nabla c} \qquad (4.30)$$

To summarize, one can say that two overall fracture conductivities can be defined according to the calculation method, namely the Laplace and the Reynolds conductivities denoted by Λ_L and Λ_R which are solutions to the systems (4.22)–(4.24) and (4.29), respectively. System (4.29) is much easier to solve than (4.22)–(4.24), but is quite approximate.

One can also define equivalent mean apertures by inverting the law (4.27c)

$$b_{cL} = \frac{\Lambda_L}{D}, \quad b_{cR} = \frac{\Lambda_R}{D} \qquad (4.31)$$

4.3.3 Estimations of the fracture conductivity Λ

This subsection intends to provide some elements in order to estimate fracture conductivity when some geometrical parameters are known.

Again you are on a desert island; what do you do? As before, use eqn 4.27c.

If one wants to go a little further, one can transpose eqn 4.18a to macroscopic conductivity, since this latter quantity only depends on geometry as fracture transmissivity

$$\Lambda'_L = \frac{\Lambda_L}{Db_m} = g\left(\frac{b_m}{\sigma_h}, \frac{\ell_c}{\sigma_h}, \theta_I, \text{fracture nature}\right) \qquad (4.32)$$

Some results are shown in Fig. 4.6 for Gaussian fractures of the type (2.8) while self-affine fractures are addressed in Section 4.5. They are obtained in the same way as transmissivity in Section 4.2.3, i.e. by statistical averaging over a sufficiently large number of independent realizations and they can be discussed in the same way too. The same jokes

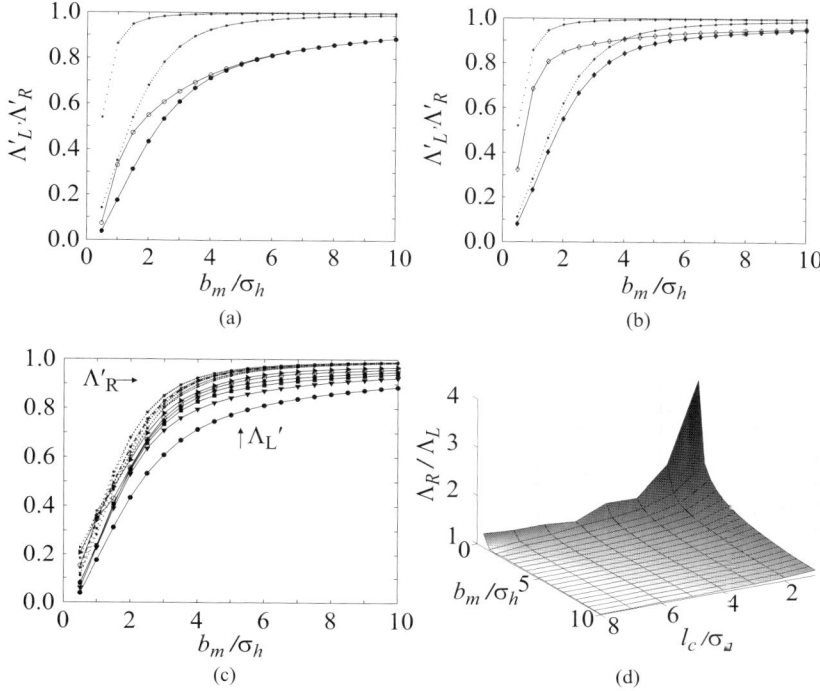

Fig. 4.6 Dimensionless conductivities Λ'_L (solid lines) and Λ'_R (broken and dotted lines) of Gaussian fractures. The values are conductivity averages over 25 independent realizations along the x- and y-axes; the size of the unit cell is extrapolated to infinity; $\sigma_h = 5a$. Same conventions as in Fig. 4.5. (a) Influence of the intercorrelation coefficient θ_I for $\ell_c/\sigma_h = 1$ (a) and 4 (b). (c) Λ'_L and Λ'_R as functions of b_m/σ_h; $\theta_I = 0$. (d) The ratio Λ_R/Λ_L as a function of b_m/σ_h and ℓ_c/σ_h for $\theta_I = 0$.

can be made! The influence of the correlation coefficient θ_I is illustrated in (a) and (b) with approximately the same conclusion as before, i.e. this influence increases with the ratio ℓ_c/σ_h. The same figures (a) and (b) show that the Reynolds approximation depends on θ_I, and not on ℓ_c/σ_h with the same explanation as previously.

The Reynolds approximation is not all that valid and is an overestimation of the true conductivity Λ'_L; but nevertheless, this overestimation is smaller than that of transmissivity as discussed in Section 4.3.4.

Systematic results for Λ'_L and Λ'_R are provided in Fig. 4.6c for uncorrelated fracture surfaces. Λ'_L increases with b_m/σ_h and with ℓ_c/σ_h and these variations can be interpreted in terms of roughness. ℓ_c/σ_h has a significant influence on Λ'_L. Again Λ'_L is seen to tend towards Λ'_R when ℓ_c/σ_h increases.

4.3.4 Comparison between Reynolds and Laplace conductivities

For the same reason as for transmissivity, let us now devote a short subsection to this important topic.

The numerical results obtained for conductivity by the two methods are compared in Fig. 4.6d. Once again, it is for the range of the experimental data recalled by (4.19) that the ratio Λ_R/Λ_L is the largest. Therefore, it is important to use the results derived from the exact Laplace equation even though it is more difficult to solve than the Reynolds approximation.

It is also interesting to compare Fig. 4.5c with Fig. 4.6d. The discrepancy between the Reynolds approximation and the exact calculations is smaller for conductivity than for transmissivity. Of course, this is due to the fact that conductivity is proportional to the aperture b while transmissivity is proportional to the cube of the aperture b^3, and hence there is a greater discrepancy.

4.4 Dispersion of a passive solute

The study of this phenomenon is much less advanced than the two previous ones. This section only provides a general presentation of this transport with a few numerical results.

Consider the transport of a passive Brownian solute by diffusion and convection through a fracture. Recall that the flux j of a solute is described as the sum of diffusive and convective contributions

$$j = vc - D\nabla c \tag{4.33}$$

where c is the solute volumetric concentration, v the fluid velocity and D the molecular diffusion coefficient. The velocity field obeys eqn 4.1 and should be known beforehand. Conservation of the solute yields the convection-diffusion equation

$$\frac{\partial c}{\partial t} + v \cdot \nabla c - D\nabla^2 c = 0 \tag{4.34}$$

Generally, the fracture solid boundaries $S_p^\pm$ are supposed to be impervious to the solute and not to react with it. Therefore, by using the adherence condition (4.2a) (cf. also Fig. 4.1)

$$n \cdot \nabla c = 0 \quad \text{on} \quad S_p^\pm \tag{4.35}$$

System (4.34, 4.35) is a complicated local problem. Furthermore, one is generally interested only in the moments of the solute distribution. Suppose that at time $t = 0$ a unit quantity of solute is released into the fluid. In the absence of chemical reactions, the total amount of solute is invariant

$$M_0(t) = \int_{\tau_f} c(\boldsymbol{x},t) \, \mathrm{d}^3\boldsymbol{x} = 1 \tag{4.36}$$

where τ_f is the volume of the fracture. The first moment $\boldsymbol{M}_1$ indicates the average position of the solute

$$\boldsymbol{M}_1(t) = \int_{\tau_f} \boldsymbol{x} \, c(\boldsymbol{x},t) \, \mathrm{d}^3\boldsymbol{x} \tag{4.37}$$

The centered second moment $\boldsymbol{M}_2$ describes the spreading of the solute

$$\boldsymbol{M}_2(t) = \int_{\tau_f} (\boldsymbol{x} - \boldsymbol{M}_1)(\boldsymbol{x} - \boldsymbol{M}_1) \, c(\boldsymbol{x},t) \, \mathrm{d}^3\boldsymbol{x} \tag{4.38}$$

Note that $M_0(t)$, $\boldsymbol{M}_1(t)$ and $\boldsymbol{M}_2(t)$ are a scalar, a vector and a second order tensor, respectively. Moreover, we are not interested in the transients which occur just after the introduction of the solute into the fluid, but rather in the asymptotic regime which develops at long times (for more details, see for instance Adler, 1992). In this limit, $\boldsymbol{M}_1$ and $\boldsymbol{M}_2$ verify

$$\lim_{t \to \infty} \frac{d\boldsymbol{M}_1}{dt} = \overline{\boldsymbol{v}}^* \qquad \lim_{t \to \infty} \frac{1}{2}\frac{d\boldsymbol{M}_2}{dt} = \boldsymbol{D}^* \qquad (4.39\text{a})$$

where $\overline{\boldsymbol{v}}^*$ is the mean interstitial fluid velocity defined as the spatial average of the fluid velocity over the fluid space τ

$$\overline{\boldsymbol{v}}^* = \frac{1}{\tau} \int\!\!\int\!\!\int_\tau \boldsymbol{v}\, d^3\boldsymbol{x} \qquad (4.39\text{b})$$

$\boldsymbol{D}^*$ is the dispersion tensor. For applications, $\overline{\boldsymbol{v}}^*$ and $\boldsymbol{D}^*$ are the two important quantities, since they describe the mean displacement and the spreading of the solute, respectively. They appear as effective coefficients in an upscaled convection-dispersion equation with the same form as eqn 4.34

$$\overline{b}\frac{\partial \overline{c}}{\partial t} + \nabla_X \cdot \left[\overline{b}(\overline{\boldsymbol{v}}^*\overline{c} - \boldsymbol{D}^* \cdot \nabla_X \overline{c})\right] = 0 \qquad (4.40)$$

where $\overline{b}$ is a local average of the aperture. ∇_X is a two-dimensional gradient operator on a scale large with respect to $\overline{b}$.

Equation 4.34 can be made dimensionless by choosing basic units V and b_0 for velocity and length; the latter is usually a characteristic fracture aperture. The Péclet number appears naturally in (4.34) and is expressed as

$$Pe = \frac{V b_0}{D} \qquad (4.41)$$

$\boldsymbol{D}^*$ depends on the geometry of the fracture, but also on the Péclet number. An estimate of the Péclet number in realistic situations can be obtained as follows. Since a typical diffusion coefficient for salts in water is $D = 10^{-9} \text{m}^2/\text{s}$, Pe is of order 1 if the mean fluid velocity $\overline{v}^*$ is of the order of 1 meter per day ($\sim 10\mu\text{m/s}$) and if the fracture aperture b is in the range $0.1 \sim 1\text{mm}$. Hence, a reasonable range for Pe appears to be

$$10^{-2} \leq Pe \leq 10^2 \qquad (4.42)$$

For a plane Poiseuille flow (cf. Fig. 4.2), the dispersion tensor has only two independent components $D^*_\parallel$ and $D^*_\perp$

$$\boldsymbol{D}^* = \begin{pmatrix} D^*_\parallel & 0 & 0 \\ 0 & D^*_\perp & 0 \\ 0 & 0 & 0 \end{pmatrix} \qquad (4.43)$$

Wooding (1960) derived the corresponding result between two flat plates separated by a distance b_m

$$D_\parallel^* = D\left(1 + \frac{Pe^2}{210}\right) \qquad D_\perp^* = D \qquad Pe = \frac{\overline{v}^* b_m}{D} \qquad (4.44)$$

Data relative to transport through random fractures with Gaussian correlations are presented in Fig. 4.7. The mean aperture b_m and the correlation length ℓ_c vary from σ_h to $3\sigma_h$ and from $2\sigma_h$ to $4\sigma_h$, respectively. The unit cell size of the periodic fracture is $20\sigma_h$. The influence of the mean aperture is illustrated in Fig. 4.7a, for $\ell_c = 2\sigma_h$. Transition between diffusive and dispersive regimes is smoother than in a plane channel, and occurs for a smaller Péclet number. For all apertures, $D_\parallel^*$ and $D_\perp^*$ can be approximated by power laws of the form Pe^α when $Pe > 1$, where α is smaller than 2. When measured between $Pe = 1$ and $Pe = 100$, α is about 1.3. $D_\parallel^*$ increases when b_m decreases, as a consequence of larger in-plane relative variations on the fluid flux and of increasing fractional contact area ($S_c = 0.02, 0.10, 0.18$ and 0.29 for $b_m/\sigma_h = 3, 2, 1.5$ and 1, respectively). Surprisingly, the correlation length ℓ_c which governs the size of the contact areas and the width of the main flow paths has little influence (Fig. 4.7b).

A more general analysis of solute dispersion in fractures should take into account the existence of multiple length scales on which velocity fluctuations take place. The smallest one is the aperture b across which the velocity profile is approximately parabolic (see Fig. 4.2). On a larger scale of the order of the correlation length ℓ_c, flow channelization can occur because the fluid has to circumvent contact zones or low aperture regions. This can induce the coexistence of near-stagnant and fast-flowing zones. Finally, regional variations in the fracture characteristics can exist at an even larger scale $\mathcal{L}$. The establishment of the asymptotic regime eqn 4.39a results from the diffusive exchanges between streamlines, and the associated characteristic times required to sample these velocity fluctuations are b^2/D, ℓ_c^2/D and $\mathcal{L}^2/D$, respectively. Péclet numbers associated with these various scales can be defined similarly to eqn 4.41, and different dispersion regimes and dispersion coefficients can prevail according to the scale of observation and to the value of the corresponding Péclet number. Note that in the example in Fig. 4.7, b_m and ℓ_c are of similar orders of magnitude, and the fractures are periodic so

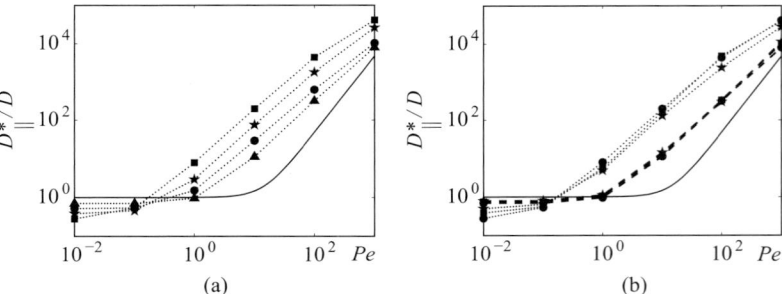

Fig. 4.7 Longitudinal dispersion coefficient $D_\parallel^*/D$ versus the Péclet number in random fractures with Gaussian correlations. (a) $\ell_c = 2\sigma_h$ and $b_m/\sigma_h = 1$ (■), 1.5 (⋆), 2 (•) and 3 (▲). (b) $b_m/\sigma_h = 1$ (·····) and 3 (– – –); $\ell_c/\sigma_h = 2$ (•), 3 (■) and 4 (⋆). The solid line is eqn 4.44.

that there are no regional variations in fracture properties. Therefore, this multiscale aspect is not important.

4.5 Extensions

Three topics are addressed in this section. Systematic calculations made in the framework of the lubrication approximation for wavy channels are first presented and transmissivity is given by (4.48). Results for the conductivity and the transmissivity of self-affine fractures are provided in Section 4.5.2 by eqns 4.52 and 4.53, respectively.

4.5.1 Plane channels with wavy walls

Malevich *et al.* (2006) solved Stokes flow through channels enclosed by two wavy walls whose amplitude is proportional to the mean aperture of the channel multiplied by the small dimensionless parameter ε. The application of an analytical-numerical algorithm yields efficient formulas for the velocities and transmissivity. These formulas include ε in symbolic form.

An example is provided by the two-dimensional channel bounded by the surfaces

$$z^+ = \frac{b_m}{2}(1 + \varepsilon \cos x), \quad z^- = -\frac{b_m}{2}(1 + \varepsilon \cos x). \quad (4.45)$$

The dimensionless transmissivity $\sigma'(\varepsilon)$ is defined according to (4.18) and is calculated up to $O(\varepsilon^{32})$

$$\sigma'(\varepsilon) = \frac{12}{b_m^3}\sigma(\varepsilon) = 1 - \sum_{n=1}^{\infty} c_{2n}\varepsilon^{2n} = \sigma_{30}(\varepsilon) + O(\varepsilon^{32})$$
$$= 1 - 3.14963\varepsilon^2 + 4.08109\varepsilon^4$$
$$- 3.48479\varepsilon^6 + 2.93797\varepsilon^8 - 2.56771\varepsilon^{10} + 2.21983\varepsilon^{12} - 1.93018\varepsilon^{14}$$
$$+ 1.67294\varepsilon^{16} - 1.45302\varepsilon^{18} + 1.26017\varepsilon^{20} - 1.09411\varepsilon^{22} + 0.949113\varepsilon^{24}$$
$$- 0.823912\varepsilon^{26} + 0.714804\varepsilon^{28} - 0.620463\varepsilon^{30} + O(\varepsilon^{32})$$
$$(4.46)$$

The general advantages of this expansion are clear. First, this analytical expression is valid for a wide range of values of ε; if one wishes to obtain transmissivity with a precision equal to 10^{-3}, the previous expression is valid up to $\varepsilon = 0.8$. Second, the flow field is itself obtained analytically with the same precision.

The present results were successfully compared with the results obtained by a finite difference code. An analytical formula such as eqn 4.46 possesses a specific bonus. The coefficients c_{2n} in eqn 4.46 for $n \geqslant 4$ verify

$$c_{2n} = C_0(-q)^{-n} \quad (4.47)$$

with $C_0 = 5.19679$ and $q = 1.15220$ with very high precision. By using this rule and the condition $\sigma'(1) = 0$, we can extend eqn 4.46 by adding

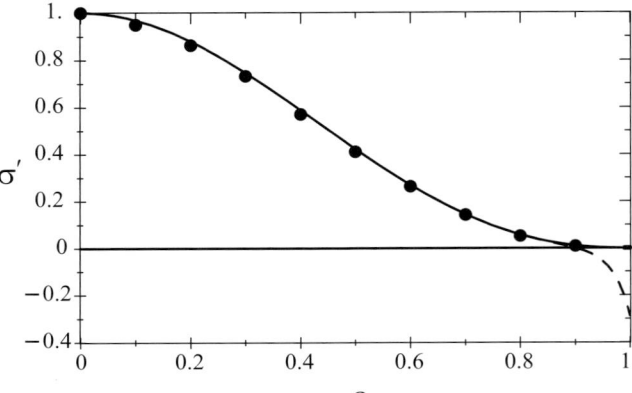

Fig. 4.8 The normalized transmissivity σ' as a function of ε for the channel defined by eqn 4.45. Data are for: solid line (4.48), broken line (4.46), dots (numerical solution).

an infinite series of terms of the form eqn 4.47. Then, the full expansion for transmissivity can be summed and we obtain

$$\sigma'(\varepsilon) = \sigma'_{30}(\varepsilon) + \frac{\alpha\, \varepsilon^{32}}{\beta + \varepsilon^2}, \qquad (4.48)$$

where $\alpha = 0.604220$, $\beta = 1.09880$ are again calculated via the least square method. Functions $\sigma'(\varepsilon)$ and $\sigma'_{30}(\varepsilon)$ are displayed in Fig. 4.8, where they are seen to superpose exactly.

4.5.2 Extension to self-affine fractures

Conductivity

The fracture surfaces are generated as described in Subsection 2.6.2 with eqn 2.43. To avoid repetition, the reader is invited to read this subsection again where all the concepts and notations are introduced, including the necessary distinction between spatial and statistical averages (see Section 2.6.2). Square fracture samples of size $L \times L$ are reconstructed, by generating heights h^+ and h^- at the nodes $(i\Delta x, j\Delta y)$ of a regular square grid with $\Delta x = \Delta y = a$, $L = \mathcal{N}_c a$ and $i, j = 1, 2, ..., \mathcal{N}_c$. Then, the $L \times L$ fracture sample F is split into a collection of square domains Ω_λ of varying size $\lambda = n_\lambda a$ whose conductivities are determined. The boundary conditions are provided by eqn 4.24 where L should be replaced by λ.

The boundary conditions (4.24) are chosen since they are those which would be most easily imposed in a real experiment; a piece of fracture can be cut off and wrapped to prevent any leakage. The problem (4.22, 4.23, 4.24) is solved as described by Mourzenko *et al.* (1999).

The average flux $\overline{\boldsymbol{J}}$ per unit fracture width can be defined as

$$\overline{\boldsymbol{J}} = -\frac{1}{\lambda^2} \int_{\tau_f} D \nabla c \, \mathrm{d}^3 \boldsymbol{x} \qquad (4.49)$$

Such a definition is more general than eqn 4.25. Moreover, the overbar corresponds to the notations eqn 2.45. Dimensionless conductivity $\overline{\Lambda}_x$ is deduced from the x-component $\overline{J}_x$ of $\overline{\boldsymbol{J}}$ by

$$\overline{J}_x = -\overline{\Lambda}_x \frac{c_2 - c_1}{\lambda} \qquad (4.50)$$

$\overline{\Lambda}_x/D$ can be seen as the aperture of an equivalent plane channel (cf. also eqns 4.27c and 4.31), which yields the same flux $\overline{J}_x$ in the same conditions (4.24).

Notations for spatial averages and variances over subdomains Ω of F with size λ and for conditional averages of $\overline{X}$ over domains which share a common value y of some parameter Y are provided by eqns 2.45 and 2.46.

The statistical averages of the local dimensionless conductivities averaged over domains with identical open areas $S_{0,\lambda}$ can be combined in a single and very important empirical relationship (see Fig. 4.9)

$$\langle \overline{\Lambda}_x/\overline{\sigma}_b \rangle_{S_{0,\lambda}} = D S_{0,\lambda}^{5-3H} \qquad (4.51)$$

To summarize, the conductivity of finite samples of fractures in the self-affine regime, with vanishing to moderate relative apertures, can be fully described by the combination of eqn 4.51 with the scaling law (2.48) for $\overline{\sigma}_b$

$$\overline{\Lambda}_x = D\mathcal{C}_\Lambda \, \lambda^{H_b} \, S_{0,\lambda}^{5.5-3H}, \qquad S_{0,\lambda} \leq 0.8 \text{ or } \overline{b}/\overline{\sigma}_b \leq 1.2 \qquad (4.52)$$

where H_b depends slightly on b_m/σ_h, but remains close to H. $S_{0,\lambda}$ does not depend on the sample size, in the average. It is related to the mean relative aperture $\overline{b}/\overline{\sigma}_b$ by (2.49).

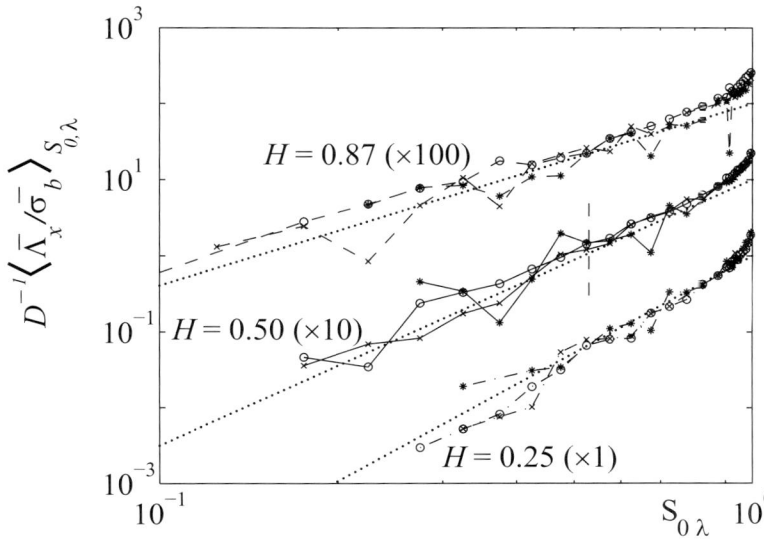

Fig. 4.9 Log-log plot of reduced average conductivity $D^{-1}\langle \overline{\Lambda}_x/\overline{\sigma}_b \rangle_{S_{0,\lambda}}$ for self-affine fractures with $H = 0.87$ (— — —), 0.50 (———) and 0.25 (— · — · —), and $n_\lambda = 32$ (○), 64 (×) or 128 (⋆), versus $S_{0,\lambda}$. For the sake of clarity, the data for $H = 0.5$ and 0.87 were shifted by one and two decades vertically, respectively. The dotted lines are the model (4.51). The vertical broken line corresponds to the percolation transition for $H=0.5$ from Mourzenko et al. (1996a).

The constant $\mathcal{C}_\Lambda$ can be deduced from measurements of the conductivity on a finite domain.

Transmissivity

A similar analysis was performed for transmissivity by Mourzenko *et al.* (2001). Note that Madadi and Sahimi (2003) studied the same problem by means of a Lattice Boltzmann algorithm. The hydraulic aperture $\overline{b}_S$, which has the dimension of length, is defined by (4.17). It can be viewed as the aperture of an equivalent plane channel, which yields the same flow rate under the same pressure gradient.

Data for $\langle \overline{b}_s/\overline{\sigma}_b \rangle_{S_{0,\lambda}}$ are plotted against the fractional open area $S_{0,\lambda}$ in Fig. 4.10. For the sake of clarity, the data for $H = 0.5$ and 0.87 were shifted by one and two decades vertically, respectively. It again appears that the data for various domain sizes are well combined in this representation, which means that the scale effects on transmissivity are fully accounted for by the normalization of the hydraulic aperture by the local aperture standard deviation $\overline{\sigma}_b$. The data for different n_λ gather fairly well around straight lines without any adjustable parameter. Least square fits of the data for $n_\lambda = 96$ yield

$$\langle \overline{b}_s/\overline{\sigma}_b \rangle_{S_{0,\lambda}} = 1.80 \; S_{0,\lambda}^{4.34} \qquad H = 0.25 \qquad (4.53\text{a})$$
$$\langle \overline{b}_s/\overline{\sigma}_b \rangle_{S_{0,\lambda}} = 1.96 \; S_{0,\lambda}^{3.52} \qquad H = 0.50 \qquad (4.53\text{b})$$
$$\langle \overline{b}_s/\overline{\sigma}_b \rangle_{S_{0,\lambda}} = 1.77 \; S_{0,\lambda}^{2.58} \qquad H = 0.87 \qquad (4.53\text{c})$$

The exponents in eqn 4.53 are very close to $5 - 3H$.

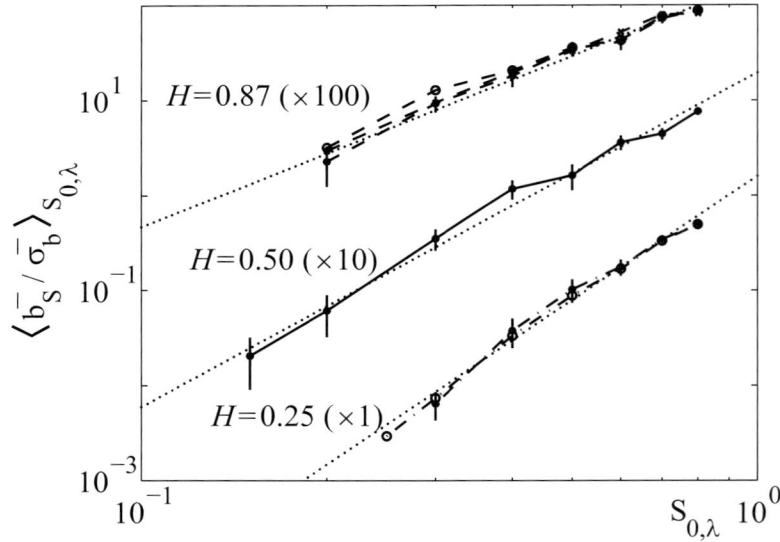

Fig. 4.10 Log-log plot of the reduced hydraulic aperture $\langle \overline{b}_S/\overline{\sigma}_b \rangle_{S_{0,\lambda}}$ for self-affine fractures with $H = 0.87$ (- - - -), 0.50 (———) and 0.25 (- · - · -), and $n_\lambda = 32$ (o), 64 (×) or 96 (⋆), versus $S_{0,\lambda}$. For clarity's sake, the data for $H = 0.5$ and 0.87 were shifted by one and two decades vertically, respectively. The dotted lines correspond to the power law fits (4.53). Vertical lines are 95% confidence intervals.

Exercises

(4.1) (i) Consider a plane rectangular channel with a constant aperture $b_m = 10^{-4}$ m, a width W equal to 2 m and submitted to a pressure gradient of 10 cm of water per meter. Determine the flow rate Q which flows through this slit.

(ii) Consider a rough fracture with an average aperture $b_m = 10^{-4}$ m, a roughness $\sigma_h = 2.10^{-5}$ m, a correlation length $\ell_c = 6.10^{-5}$ m; the two surfaces are independent. The width is equal to 2 m. The fracture is submitted to a pressure gradient of 10 cm of water per meter. Determine the flow rate Q through this rough fracture.

5 Transport in fracture networks

5.1 Introduction 86
5.2 General 87
5.3 Numerical methodology 91
5.4 I^{2}OUD fracture networks 94
5.5 Extensions 102
Exercises 108

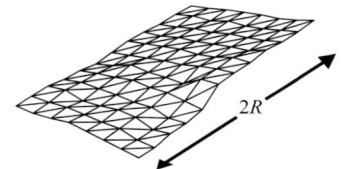

Fig. 5.1 A single fracture viewed from far which is not plane and which is triangulated.

5.1 Introduction

Let us now revert, and for the rest of this book, to a scale of the order of the lateral extension $2R$ of the fractures. At this scale, the fractures look like a single surface (cf. also Section 3.1).

A *fracture network* is a set of fractures which may intersect or not. These surfaces are relatively plane, but this is not a condition for the computational codes since they only need to be triangulated. By relatively plane, we mean surfaces whose radii of curvature are very large with regard to the average aperture $\bar{b}$. An example of such a non-planar surface is given in Fig. 5.1.

At each triangle of the fracture, one can assign a fracture transmissivity σ. More precisely, the transmissivity of each triangle of the fracture is known.

There is also a point in our terminology which needs to be recalled from Section 1.2. In a *fracture network*, the solid matrix located between the fractures is assumed to be impermeable.

The terminology of this chapter is that of flow and permeability. However, on the Darcy scale the same equations govern phenomena such as diffusion which obey the Laplace equation on the pore scale (cf. Section 4.3), hence the title of this chapter.

This chapter is organized as follows. Section 5.2 is devoted to the presentation of the equations which govern flow in fracture networks; the elegant solution due to Snow (1969) for infinite fractures is derived. Certain details concerning meshing are provided in Section 5.3 as well as some elementary examples and also some considerations regarding the actual possibilities of the mesher and of the flow solver. Numerical data for the basic I^{2}OUD case where fractures have the same shape and size, and are isotropically oriented and uniformly distributed are gathered together in Section 5.4 where their physical significance is also discussed.

Just as in the previous chapters, Section 5.5 generalizes the previous results to more complex structures where the sizes of the fractures obey a power law or when they are no longer isotropic.

5.2 General

Most of this section is a summary of a paper by Koudina *et al.* (1998).

5.2.1 Flow equations

The solid matrix containing the fractures is assumed to be impervious. The flow of a Newtonian fluid at low Reynolds number is governed by the Stokes equations within a fracture, i.e. at a local scale characterized by the average aperture $\bar{b}$. On a scale large with respect to $\bar{b}$ (cf. eqn 4.4), the flow is governed by the Darcy equation (4.4b)

$$\boldsymbol{J} = -\frac{1}{\mu} \sigma \nabla_s p \qquad (5.1)$$

$\boldsymbol{J}$ and $\nabla_s p$ are the locally averaged flow rate per unit width $[L^2 T^{-1}]$ and the pressure gradient, respectively; σ $[L^3]$ is the local fracture transmissivity.

For simplicity sake, fracture transmissivity is assumed to be isotropic in the fracture plane and is denoted by the scalar σ. The flow rate per unit width $\boldsymbol{J}$ belongs to the fracture plane since the embedding solid is impermeable. The 2D gradient operator ∇_s works in the fracture plane where pressure is defined. Mass conservation implies the 2D continuity equation via reasoning which is similar to that which provides (4.13)

$$\nabla_s \cdot \boldsymbol{J} = 0 \quad \text{or equivalently} \quad \nabla_s \cdot (\sigma \nabla_s p) = 0 \qquad (5.2)$$

σ may depend on the fracture and on the position of the point inside the fracture. Note that the fracture intersections could be given specific transmissivities (Garcia-Gonzales *et al.*, 1999).

These equations have to be supplemented by two classes of boundary conditions on each fracture which are illustrated in Fig. 5.2a. First, at the external boundary $\mathcal{L}_f$ of the fractures, the flow which is normal to the boundary should vanish since the solid medium located in between the fractures is impervious

$$\boldsymbol{n}_{eb} \cdot \nabla_s p = 0 \qquad (5.3a)$$

where $\boldsymbol{n}_{eb}$ is the normal to $\mathcal{L}_f$ in the fracture plane. Second, along an intersection I_{ij} between the two fractures i and j, the pressures p_i and p_j are equal, and the total flux J_{ij} normal to I_{ij} which arises from all the intersecting fractures is equal to zero. More precisely,

$$p_i = p_j, \ J_{ij} = 0 \quad \text{on} \quad I_{ij} \qquad (5.3b)$$

This set of boundary conditions is incomplete. External boundary conditions, including driving force, are detailed in the next subsection.

5.2.2 Permeability K_n of a fracture network

The basic flow experiment to determine permeability of the fracture network is designed according to Fig. 1.4 and is applied to fracture networks as illustrated in Fig. 5.2b. Take a piece of fractured medium

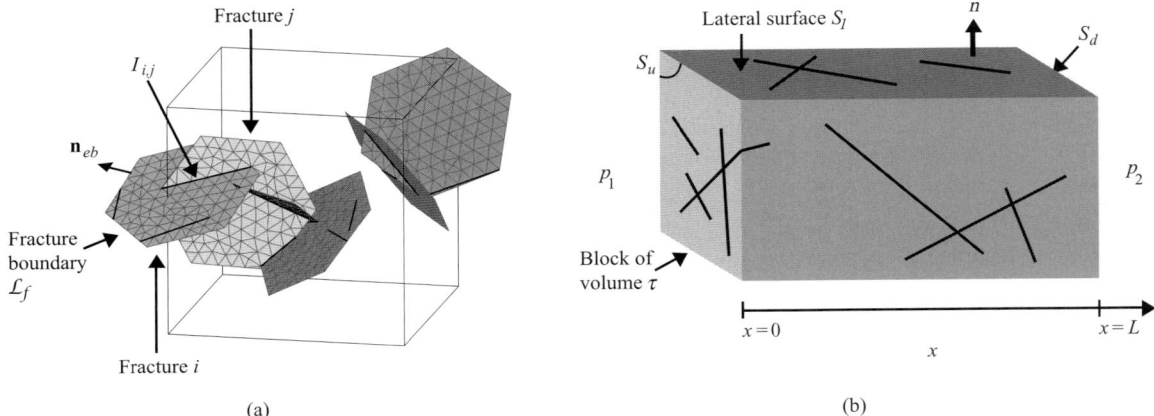

Fig. 5.2 Boundary conditions on a fracture network. (a) Conditions on each fracture. (b) Overall conditions.

of volume τ, wrap it in an impermeable cover and put it between two vessels with two pressures p_1 and p_2 applied to the upstream and downstream surfaces S_u and S_d.

Therefore, one has to add the two overall boundary conditions to the equations in Section 5.2.1

$$p(x=0) = p_1 \quad \text{on} \quad S_u, \quad p(x=L) = p_2 \quad \text{on} \quad S_d; \quad \boldsymbol{J} \cdot \boldsymbol{n} = 0 \quad \text{on} \quad S_l \tag{5.4}$$

The component $\overline{\overline{v}}_x$ of the seepage velocity along the x-axis is obtained by averaging local fluxes over the fracture surfaces. Since the equations are linear, every quantity is proportional to the overall pressure gradient $\overline{\overline{\frac{\partial p}{\partial x}}}$

$$\overline{\overline{v}}_x = \frac{1}{\tau} \int_{\sum_j S_j} J_x \mathrm{d}S = -\frac{K_n}{\mu} \overline{\overline{\frac{\partial p}{\partial x}}} \tag{5.5}$$

where S_j is the surface of fracture j contained in the volume τ and $\overline{\overline{\frac{\partial p}{\partial x}}} = (p_2 - p_1)/L$. The double bar indicates that the average is taken over the whole block displayed in Fig. 5.2 according to the convention defined in Section 2.2.1. K_n is the network permeability.

A few remarks need to be made. K_n is homogeneous to the square of a length since the integration in (5.5) is over surfaces, and since the result is divided by a volume. K_n only depends on the geometry of the fracture network. A dimensionless permeability K_n' can be defined as

$$K_n' = \frac{K_n R_o}{\sigma_o} \tag{5.6}$$

where R_o and σ_o are the characteristic fracture size and transmissivity.

To summarize, eqn 5.2 should be solved with the boundary conditions (5.3) on each fracture and the overall boundary conditions (5.4). The network permeability K_n is deduced from Darcy's law (5.5). If the fracture network is not isotropic, the network permeability is the tensor $\boldsymbol{K}_n$.

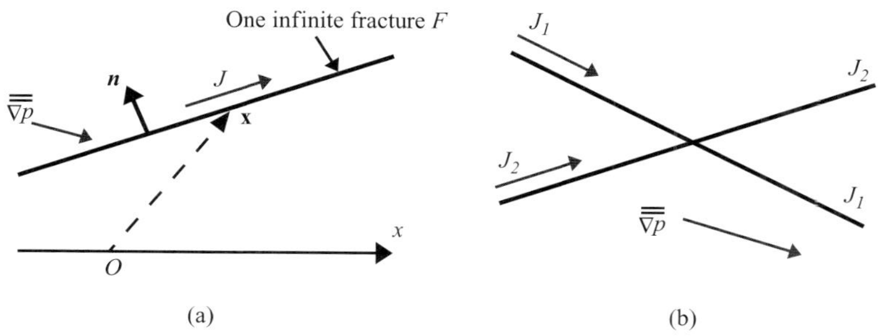

Fig. 5.3 The fundamental property for infinite fractures which are represented as thick lines. (a) One infinite fracture submitted to a macroscopic pressure gradient $\overline{\overline{\nabla p}}$. (b) Two infinite fractures submitted to $\overline{\overline{\nabla p}}$.

5.2.3 Flow through networks made of infinite fractures

This subsection is based on the very elegant solution of Snow (1969) who derived the permeability of fracture networks made of infinite planes.

Let us first consider a single infinite fracture $\mathcal{F}$ with a constant aperture, whose unit normal is $\boldsymbol{n}$; it is filled with a fluid of viscosity μ. The projection operator $(\boldsymbol{I} - \boldsymbol{nn})$ projects any vector onto the plane perpendicular to $\boldsymbol{n}$ (cf. Exercise 5.1); $\boldsymbol{I}$ is the unit tensor and $\boldsymbol{nn}$ the tensorial product of $\boldsymbol{n}$ by itself. Assume that an arbitrary pressure gradient $\overline{\overline{\nabla p}}$ is imposed. The component of the pressure gradient which lies in the fracture plane is $(\boldsymbol{I} - \boldsymbol{nn}) \cdot \overline{\overline{\nabla p}}$. The resulting flux may be expressed as

$$\boldsymbol{J} = -\frac{\sigma}{\mu}(\boldsymbol{I} - \boldsymbol{nn}) \cdot \overline{\overline{\nabla p}} \qquad (5.7a)$$

where σ is equal to the constant fracture transmissivity. The pressure inside the fracture is trivially expressed through the overall pressure gradient

$$p = \overline{\overline{\nabla p}} \cdot \boldsymbol{x} \qquad (5.7b)$$

where $\boldsymbol{x}$ is the spatial position as illustrated in Fig. 5.3a.

It is easy to realize that a set of two infinite plane fractures which intersect one another does not introduce any additional complexity. This is illustrated in Fig. 5.3b. There is a trivial solution which consists in assuming that the pressure in both fractures is provided by eqn 5.7b. Then, the necessary equality of pressures at the intersection is automatically verified and the first part of eqn 5.3b is satisfied. If the fluxes are conserved at the intersection as shown in Fig. 5.3b, the condition on the flux conservation at the intersection is satisfied. Therefore, one has a trivial solution to the problem involving two fractures. When one has a solution to this sort of linear problem, it is *the solution*. It should be noted that the two fractures may have different transmissivities provided that they are constant over each fracture.

The total flux in Fig. 5.3b is simply

$$\boldsymbol{J} = \boldsymbol{J}_1 + \boldsymbol{J}_2 \qquad (5.8)$$

This last property can easily be generalized to a set of infinite fractures j of surface transmissivity σ_j. Consider a volume τ containing the surface

S_j of each fracture of the set. The seepage velocity can be expressed as (cf. eqn 5.7)

$$\overline{\boldsymbol{v}} = \frac{1}{\tau} \int_{\sum S_j} \boldsymbol{J}_j \mathrm{d}S = \frac{1}{\tau} \sum_j \boldsymbol{J}_j S_j = \frac{1}{\tau} \sum_j \left[-\frac{1}{\mu} \sigma_j (\boldsymbol{I} - \boldsymbol{n}_j \boldsymbol{n}_j). \overline{\nabla p} \right] S_j \quad (5.9\mathrm{a})$$

Therefore, (5.5) implies that the permeability $\boldsymbol{K}_{nS}$ of a network of infinite fractures is

$$\boldsymbol{K}_{nS} = \sum_j \sigma_j \mathcal{S}(\boldsymbol{n}_j)(\boldsymbol{I} - \boldsymbol{n}_j \boldsymbol{n}_j) \quad (5.9\mathrm{b})$$

where $\mathcal{S}(\boldsymbol{n}_j)$ is the surface area per unit volume of the family S_j. The subscript S stands for Snow.

These formulae can be generalized to continuous distributions of fracture orientations by replacing the sums by adequate integrals. For an isotropic distribution of fractures uniformly distributed with the same fracture transmissivity σ, Adler and Thovert (1999) showed that

$$\boldsymbol{K}_{nSr} = K_{nSr} \boldsymbol{I} \quad \text{with} \quad K_{nSr} = \frac{2}{3} \mathcal{S} \sigma \quad (5.10)$$

where $\mathcal{S}$ is the total fracture surface per unit volume. Note that the scalar network permeability K_{nSr} is indeed homogeneous to the square of a length. Here, the subscript r refers to the reference case since the network is I^{2}OUD according to the notation introduced in Section 3.4.1.

Finally, consider an I^{2}OUD network of finite fractures. Let A and P be the area and the perimeter of such fractures. If the fracture density is ρ, the fracture surface $\mathcal{S}$ per unit volume is

$$\mathcal{S} = \rho A \quad (5.11\mathrm{a})$$

which can be expressed in terms of the dimensionless density ρ' defined by (3.6)

$$\mathcal{S} = \frac{2\rho'}{P} \quad (5.11\mathrm{b})$$

The permeability of a network of infinite fractures with the same $\mathcal{S}$ can be expressed in various equivalent ways

$$K_{nSr} = \frac{2}{3} \rho \sigma A = \frac{4}{3} \sigma \frac{\rho'}{P} \quad (5.11\mathrm{c})$$

or in dimensionless terms

$$K'_{nSr} = \frac{K_n R}{\sigma} = \frac{4}{3} \frac{\rho' R}{P} = k_S \rho' \quad \text{with} \quad k_S = \frac{4R}{3P} \quad (5.11\mathrm{d})$$

The values of k_S for the fracture shapes considered here are provided in Table 5.1.

This last formula can be specialized for rectangles with an infinite aspect ratio

Table 5.1 Hydrodynamic parameters associated with disks and with the investigated fracture shapes for I^{2}OUD fractures. The corresponding geometrical parameters are given in Table 3.1. The general definition of K'_0 is given by eqn 5.17. The fits are detailed in Section 5.4.1; the overall fit corresponds to eqns 5.16b and 5.17.

Shape	Theory		Fits			
	k_S (5.11d)	K'_0 (5.23b)	α	β	k_S	K'_0
Disks	$2/(3\pi)$	$4/3$				
Hexagons	$2/9$	$8\sqrt{3}/9$	0.0406	0.180	0.226	1.77
Squares	$\sqrt{2}/6$	$4\sqrt{2}/3$	0.0359	0.148	0.243	2.21
Triangles	$4\sqrt{3}/27$	$8/3$	0.0355	0.138	0.257	2.44
6-rectangles	$\sqrt{37}/21$	$7\sqrt{37}/9$				
Overall fit			0.037	0.155	0.239	2.09

$$K'_{nSr} = \frac{\rho'}{3} \qquad (5.11e)$$

A general feature of these formulae is that the network permeability is proportional to the network density. Moreover, for a given shape, the dimensionless permeability only depends on the dimensionless density. Note in Table 5.1 that the variations of k_S are very limited since k_S ranges from 0.21 for disks to 0.33 for rectangles with an infinite aspect ratio.

5.3 Numerical methodology

Koudina *et al.* (1998) devised numerical tools for arbitrary fracture networks and arbitrary distributions of fracture transmissivities. These tools which are very versatile, comprise four steps that are going to be briefly summarized.

5.3.1 Meshing

The fractures are meshed one by one as illustrated in Fig. 5.4. A maximal discretization length δ_M is chosen; ideally, it should be small with regard to any characteristic size of the fractures. First, the intersections between the fractures are determined (a). Second, the border of each fracture and the intersections with the others are divided by segments of regular length δ at most equal to δ_M (b). Third, the *method of advancing front* is used; from every couple of successive points A and B along the border and the intersections (c), one tries to introduce a new point C located at distance δ from the two first ones (d); one has to check that the segments AC and BC do not cross any feature such as an intersection and an existing triangle; whenever this is not the case, triangle ABC is created and AC and BC now belong to the border from which new points can be

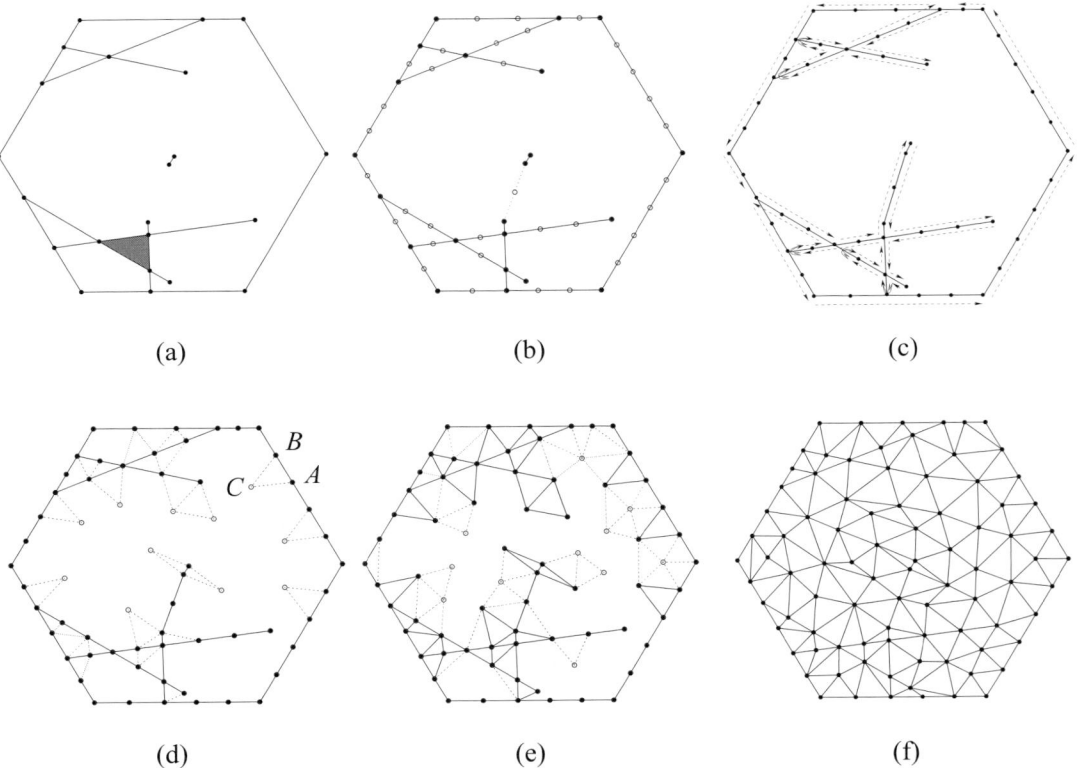

Fig. 5.4 Successive steps in the triangulation of a complex fracture. Initial geometry (a), which may include intersections lines and inner sub domains (shaded). Splitting of the edges into segments which are shorter than δ_M and connection of all internal features (b). Initial front (c); the arrows correspond to the spanning cycle. Intermediate (d, e) and final (f) stages of the triangulation.

introduced. This process is illustrated in Figs 5.4d and e; it is continued until the whole fracture is triangulated as shown in (f). Of course, during the introduction of new points, compromises have to be made and the distances AC and BC are not necessarily equal to δ_M. Additional details and adequate references are provided by Koudina *et al.* (1998).

As stated, the fractures are successively meshed. The segments which are common to two fractures and the points common to two or three fractures are labeled in a particular way. When all the fractures are meshed, the fracture network is meshed. One such example was shown in Fig. 3.1. Another example, in Fig. 5.6, will be commented on in Subsection 5.3.4.

5.3.2 Discretization of the equations and resolution

The value of pressure p must be determined at each of the N_v mesh points m of the triangular mesh (cf. Fig. 5.5). The N_v unknown pressures are determined from N_v equations, i.e. one equation per mesh point, obtained from a flux balance condition, via a first-order finite volume scheme.

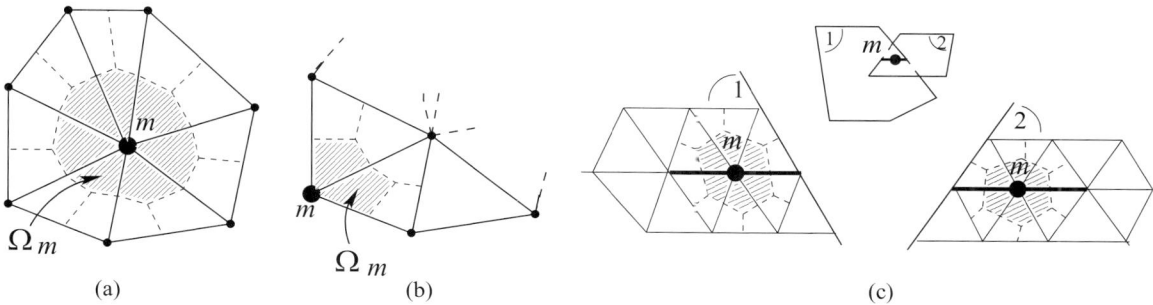

Fig. 5.5 Finite volume Ω_m surrounding a point m; neither on a fracture boundary nor on an intersection in (a); on a fracture boundary in (b); m is located on a fracture intersection denoted by the thick segment in (c). In (c), the two fractures have been separated and placed apart for the sake of clarity.

Equation 5.2 is integrated over non-overlapping domains (or control volumes) Ω_m which surround the mesh point m, as illustrated in Fig. 5.5. If m belongs to the intersection of two fractures, or even if it is a triple point (intersection of three fractures), Ω_m is simply the union of the two or three domains obtained as above in each fracture (Fig. 5.5c).

In this first-order formulation, both σ and $\nabla_s p$ are considered piecewise constant on each triangle. Of course, the driving force in these equations is the macroscopic pressure gradient $\overline{\overline{\nabla p}}$ introduced in Section 5.2.2. Finally, the discretized transport equation can be expressed as the linear system

$$\boldsymbol{A} \cdot \boldsymbol{p} - \boldsymbol{B} = 0 \qquad (5.12)$$

where $\boldsymbol{B}$ corresponds to the macroscopic pressure drop. It is easily shown that the matrix $\boldsymbol{A}$ is symmetric. Equation (5.12) is solved iteratively by a conjugate gradient algorithm. An integral convergence criterion is used

$$\|\boldsymbol{A} \cdot \boldsymbol{p} - \boldsymbol{B}\| \leq \eta \|\boldsymbol{B}\| \qquad (5.13)$$

where $\|\ \|$ denotes the standard Cartesian norm and η a small number.

5.3.3 Network permeability

Network permeability is derived from the pressure field which was determined in the previous subsection. First, seepage velocity is obtained by integration over the triangles of all the fractures according to the first equality in eqn 5.5. Second, Darcy's law is used according to the second equality in the same equations.

5.3.4 An elementary example

A *toy example* is shown in Fig. 5.6. It is a spatially periodic network which contains six fractures. The network which seems to be disconnected, is actually not so because of these periodic boundary conditions. Fracture 4 which is intersected by four other fractures, is particularly interesting and is detailed in (c) and (d). Flow is intense in the central

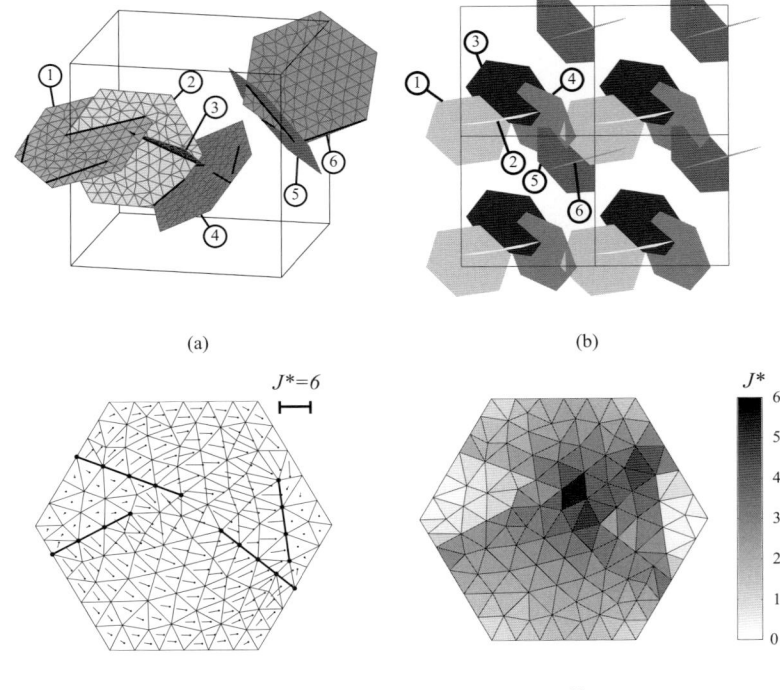

Fig. 5.6 Example of a network of six equisized regular hexagonal fractures with $L/R = 3$. (a) The unit cell. (b) A series of four adjacent unit cells viewed from above. (c) and (d) The flow field with two different representations. The segments in (c) are proportional to the local dimensionless flux $\boldsymbol{J}^* = \boldsymbol{J}\mathcal{S}/\overline{v_x}$. The color convention in (d) is from white (low velocities) to black (high velocities).

part of fracture 4 while it is relatively weak in between the intersections in the left part; flow is also very weak between the intersection at the right and the border of the fracture. One gets the impression that the fluid penetrates into fracture 4 by the two left intersections and leaves it by the two right ones.

5.4 I^{2}OUD fracture networks

5.4.1 Permeability

The first systematic calculations were performed on I^{2}OUD networks by Koudina *et al.* (1998) for relatively small densities since ρ' was limited to 16. These data were recently completed by Mourzenko *et al.* (2011b).

There is a technical point which needs to be mentioned first. The numerical results depend on the discretization length δ_M, when it is too large with respect to R. This effect was investigated first by Koudina *et al.* (1998) and then by Mourzenko *et al.* (2011b). It can be summarized by

$$K'(\delta'_M = 0) = \frac{K'(\delta'_M)}{1 + \delta'_M/\Delta'} \quad \text{with} \quad \delta'_M = \frac{\delta_M}{R} \tag{5.14a}$$

Δ' was obtained for various shapes. For instance, for hexagons, it is equal to

$$\Delta' = 0.053\, \rho' + 2.51 - 2.85/\rho' \tag{5.14b}$$

All the data which are presented in this section are extrapolated for $\delta_M = 0$.

The first crucial result is the one illustrated in Fig. 5.7. In (a), the dimensionless network permeability K'_{nr} is displayed and clearly depends to a great extent on the shape of the fractures. For instance, there are between one and two orders of magnitude between hexagons and rectangles, for the same dimensional fracture density. However, when dimensionless permeability is displayed in (b) as a function of the dimensionless density ρ', all the data seem to collapse onto the same curve. Therefore, the macroscopic network permeability K'_{nr} is not very sensitive to the fracture shape when expressed in terms of ρ'.

This is one crucial advantage of ρ'. When ρ' is used, one does not have to worry about the fracture shape. At least, not too much as we will see below.

A second feature which needs to be illustrated is the influence of the cell size L on K'_{nr}. This is shown in Fig. 5.8a. Let us introduce the convenient notation which can be easily generalized to non-I^{2}OUD networks

$$\Delta \rho' = \rho' - \rho'_{c,r} \tag{5.15}$$

where $\rho'_{c,r}$ is provided by (3.11). When $\Delta \rho'$ is less than 1, then results depend on the ratio L/R; K'_{nr} is a decreasing function of L/R. This corresponds to the finite size effect described in Section 2.5.4. When $\Delta \rho'$ increases, this effect decreases. However, one should not be misled by this representation. When plotted in arithmetic coordinates, one obtains

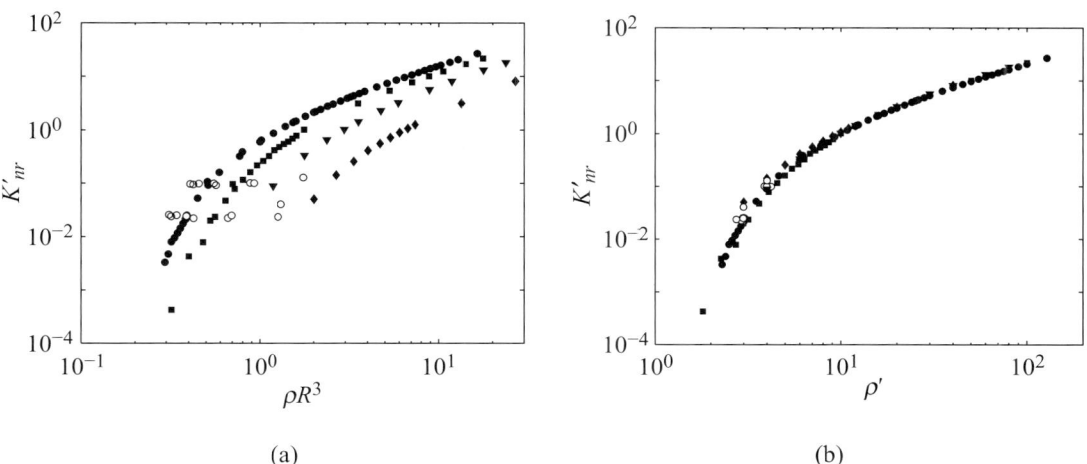

Fig. 5.7 Dimensionless permeability K'_{nr} (5.6) of I^{2}OUD fracture networks as a function of the dimensional density ρ (a) and of the dimensionless density ρ' (b). The symbols are associated with the fracture shapes: hexagons (●), squares (■), triangles (▼), 6-rectangles (♦) and others (○).

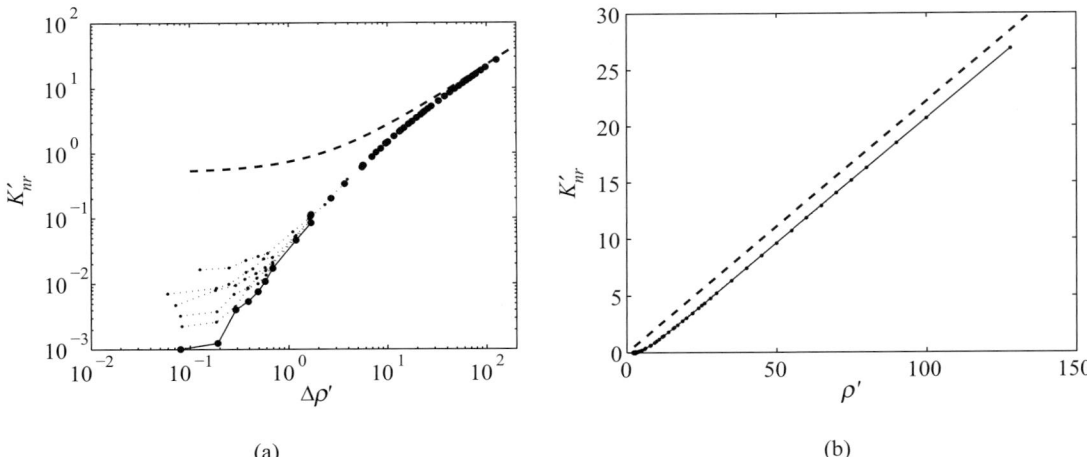

Fig. 5.8 (a) Log-log plot of the dimensionless permeability K'_{nr} of networks of I^{2}OUD hexagonal fractures as a function of $\Delta\rho' = \rho' - \rho'_{c,r}$. Dots are size-dependent data with $L/R = 4, 5, 6, 8$ and 10 (top to bottom) used to extrapolate the permeabilities for $L/R = \infty$ (•). The heavy broken line is eqn 5.11d for hexagons. (b) The same results are displayed in (b) in arithmetic coordinates.

Fig. 5.8b and the discrepancies for the various values of L/R appear to be small in a first approximation. Nevertheless, the data were extrapolated to $L/R = \infty$ as shown in Fig. 5.8a.

There is a last comment relative to Fig. 5.8a which can be made about the Snow approximation (5.11d) for hexagons. It seems clear, at least from the log-log plot, that the data for hexagons tend towards eqn 5.11d for high densities. However, Fig. 5.8b shows that there is a constant difference between the numerical data and the Snow approximation for high densities. This point will be detailed later.

The earlier sets of data for hexagonal fractures from Koudina *et al.* (1998) (for ρ' up to 12) and from Mourzenko *et al.* (2004) (for ρ' up to 20) were considerably extended by Mourzenko *et al.* (2011b) for very large densities up to $\rho' = 128$. Data were also generated for squares, equilateral triangles and 6-rectangles with ρ' up to 100, 80 and 40, respectively. Note that because of the decrease of $V_{ex,r}/R^3$, the number of fractures required to obtain a density ρ' in a sample of size L^3 strongly increases when triangles or 6-rectangles are considered. For instance, it takes about five times more 6-rectangles than hexagons to reach the same value of ρ'. For this reason, the investigated density range for the 6-rectangles is smaller than for the other shapes.

As previously noted, the calculated permeabilities are affected by finite size effects when ρ' is small and by the finite resolution δ_M of the geometrical discretization over the whole range of density. These two artifacts have been carefully characterized and corrected by Mourzenko *et al.* (2011b).

A very good fit for permeability data concerning all the fracture shapes and over the whole investigated range of density can be obtained by the formula

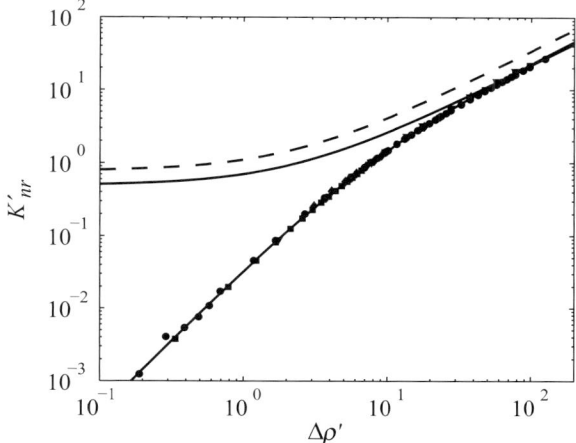

Fig. 5.9 Permeability K'_{nr} as a function of $\Delta\rho' = \rho' - \rho'_{c,r}$. The symbols represent the numerical data (same conventions as in Fig. 5.7). The thick solid curve corresponds to eqn 5.16; the Snow approximations eqn 5.11d for disks and eqn 5.11e for rectangles with an infinite aspect ratio correspond to the thin solid and the broken lines, respectively.

$$K'_{nr} = \frac{\alpha_K \Delta\rho'^2}{1 + \beta_K \Delta\rho'} \quad (5.16a)$$

with

$$\alpha_K = 0.037, \qquad \beta_K = 0.155 \quad (5.16b)$$

This functional form is heuristic, but it incorporates several features which make it very appealing. The existence of the percolation threshold is accounted for since eqn 5.16a vanishes when $\rho' = \rho'_{c,r}$. When $\Delta\rho' \ll 1$, K'_{nr} scales as $\Delta\rho'^2$, i.e. with the usual exponent $t = 2$ in percolation theory (cf. Table 2.1b).

For very large densities, K'_{nr} becomes a linear function of ρ'

$$K'_{nr} \approx k_S \rho' - K'_0, \qquad k_S = \frac{\alpha_K}{\beta_K}, \qquad K'_0 = \frac{\alpha_K (1 + \beta_K \rho'_c)}{\beta_K^2} \quad (5.17)$$

where $\rho'_{c,r}$ is provided by eqn 3.11. Since a unique pair of coefficients (α_K, β_K) is used in eqn 5.16 to represent the data for all the shapes, k_S is constant instead of shape dependent and equal to $4R/3P$ (see eqn 5.11d), but its value 0.239 is close to that for squares (0.236) and does not differ much from those for the other shapes (see Table 5.1). K'_0 slightly depends on the fracture shape through $\rho'_{c,r}$. It varies between 2.10 (squares) and 1.99 (6-rectangles).

Model (5.16) is plotted in Fig. 5.9 in comparison with the numerical results and is summarized in Table 5.1. It represents all the data within $\pm 10\%$, but the largest deviations are for very small $\Delta\rho'$ and are probably partly due to statistical fluctuations in the numerical calculations. A better agreement within $\pm 6\%$ is observed when $\rho' \geq 3$ ($\rho' \geq 6$ for the 6-rectangles).

A better precision can be achieved if the coefficients α_K and β_K in eqn 5.16a are fitted independently for each fracture shape. Fits of the data over the whole density range yield (α_K, β_K); then, (k_S, K'_0) are derived by using eqn 5.17; all these values are shown in Table 5.1. With these coefficients, eqn 5.16a represents all the data for $\rho' \geq 3$

within ±2%. They should be preferred for model networks of fractures with these particular shapes.

However, for applications to more complex cases or to field data when the geometry of the fractures is not well characterized, the robust shape-independent formula (5.16) is a very convenient and reasonably accurate all-purpose model. Note that the value of $\rho'_{c,r}$ is also nearly independent of the fracture shape, except when it strongly departs from circularity. But even then, eqn 3.11 shows that its variations are small (from 2.41 for disks to 1.90 for 6-rectangles, for instance). Therefore, a sensible choice is always possible, with little impact on the predictions of eqn 5.16 unless the network is close to critical density.

5.4.2 Properties of the velocity field

These results can be qualitatively understood in the following way. The physical situation is illustrated in Fig. 5.10 for a two-dimensional network where fractures are reduced to segments. The fracture under consideration (solid line) is intersected by a large number $N \gg 1$ of fractures (broken lines) since the dimensionless density is large. The two dotted segments belong to the fracture and are located between the two extremities of the fracture and the first (or last) intersection. Recall that from eqn 5.3a the flow should be zero at the two extremities A and D. Therefore, in two dimensions, flow is also equal to zero in the segments AB and CD. However, in the rest of the fracture between the points B and C, flow should be relatively insensitive to what happens at A and D since the end effects are expected to be masked by the large number of the intermediate intersections with the other fractures. Of course, the lengths of the two extreme segments AB and CD go to zero when fracture density increases. Therefore, in the limit of large ρ', velocity is expected to be approximately constant between B and C and close to the value expected for an infinite fracture; this explains, in a qualitative way, why network permeability tends towards the Snow formula (5.10).

This argument was made more quantitative by Mourzenko et al. (2011b); in 3D, the influence of the no-flux condition (5.3) at the border of the fracture should be less marked that in 2D, but velocity which is close to the border of the fracture should be diminished. Due to linearity of flow equations, the flow rate $\boldsymbol{J}$ at a position $\boldsymbol{x}$ in a fracture is related to the macroscopic pressure gradient $\overline{\overline{\nabla p}}$ acting over the whole network by a dimensionless matrix operator $\boldsymbol{W}$

$$\boldsymbol{J}(\boldsymbol{x}) = -\frac{\sigma}{\mu}\,\boldsymbol{W}(\boldsymbol{x}) \cdot \overline{\overline{\nabla p}} \tag{5.18}$$

The general properties of $\boldsymbol{W}$

In a network of infinite fractures, $\boldsymbol{W}$ is uniform over each fracture and only depends on the fracture orientation; it is equal to $(\boldsymbol{I} - \boldsymbol{nn})$ according to eqn 5.7a. When expressed in a referential with the z-axis which is normal to the fracture plane, it reads

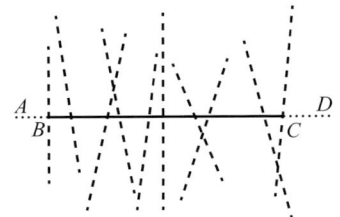

Fig. 5.10 A fracture (solid line) intersected by N fractures (broken lines); the fracture is reduced to the segment AD in 2D. The dangling ends AB and CD between the two ends and the first and last intersections are dotted; flow is zero in these dangling ends.

$$\boldsymbol{W}_\infty = \begin{pmatrix} 1 & 0 & 0 \\ 0 & 1 & 0 \\ 0 & 0 & 0 \end{pmatrix} \qquad (5.19)$$

In a network of finite fractures, $\boldsymbol{W}$ depends on the connections between the fractures, and is not uniform in each fracture. However, it is reasonable to expect that in very dense networks it is roughly given by (5.19) in the central part of the fractures.

Conversely, for a surface element near the border of a fracture which is a no-flux boundary for the flow according to (5.3a), $\boldsymbol{W}$ can be expected to be

$$\boldsymbol{W}_B \approx \begin{pmatrix} W_{B\|} & 0 & 0 \\ 0 & W_{B\perp} & 0 \\ 0 & 0 & 0 \end{pmatrix}, \quad \text{with} \quad W_{B\|} \approx 1, \quad W_{B\perp} \approx 0 \quad (5.20)$$

where the y-axis is taken normal to the border line.

The matrix $\boldsymbol{W}$ can be directly deduced from the numerical results of the flow calculations with $\overline{\overline{\nabla p}}$ successively set along three orthogonal directions. In other words, when $\boldsymbol{J}(\boldsymbol{x})$ and $\overline{\overline{\nabla p}}$ are known, $\boldsymbol{W}$ can be determined.

Attention is now focused on the trace of $\boldsymbol{W}$ which has a very important property, namely that the trace of a tensor is the same whatever the orthonormal coordinate system in which the tensor is expressed. Define

$$W = \text{trace}(\boldsymbol{W})/3, \quad W_\infty = \text{trace}(\boldsymbol{W}_\infty)/3, \quad W_B = \text{trace}(\boldsymbol{W}_B)/3$$
$$(5.21a)$$

Of course, all these values are invariant. Equations 5.19 and 5.20 imply

$$W_\infty = 2/3, \quad W_B = 1/3 \qquad (5.21b)$$

W can be normalized by W_∞

$$W' = W/W_\infty \qquad (5.21c)$$

The definitions (5.19) and (5.20) yield

$$W'_\infty = 1, \quad W'_B = 1/2 \qquad (5.21d)$$

Therefore, W' is expected to be equal to 1 and to 1/2 in the centers of the fractures and along the borders respectively.

Influence of the mesh size

The influence of the mesh size $\delta'_M = \delta_M/R$ on the results was investigated first. Results are presented in Fig. 5.11 for a hexagonal fracture taken from a network with density $\rho' = 60$. The traces of its intersections with other fractures are shown in Fig. 5.11a. The network was meshed successively with several values of δ'_M; W' in each triangular element is shown in Figs 5.11b-d which look like origamis according to R. Rosenzweig.

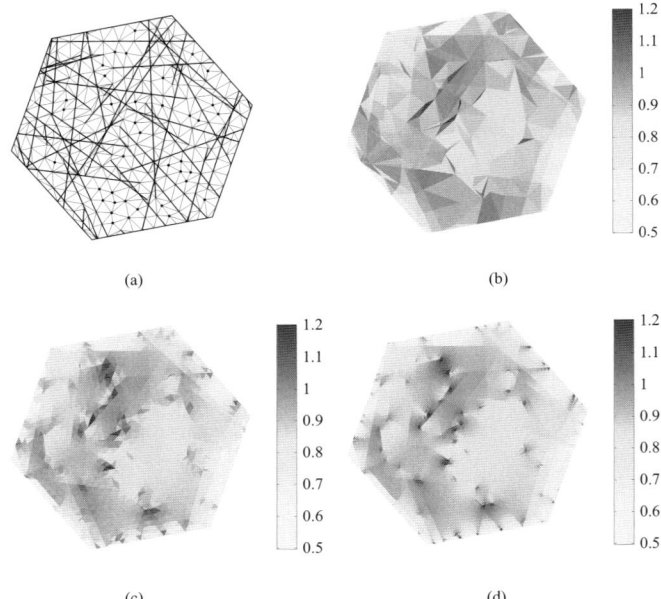

Fig. 5.11 Hexagonal fracture taken from a I²OUD network with $\rho' = 60$; in (a) are shown the intersection traces (thick lines) and the triangular mesh (thin lines) for $\delta'_M = 1/8$. The fields W' are measured in the same fracture of the same network meshed with $\delta'_M = 1/4$ (b), 1/16 (c) and 1/32 (d).

The general aspects of those three fields are similar. They comply with the expected trend with values of W' which are greater in the central part than in the peripheral region. Another interesting feature is that the peripheral region where W' is small is much more apparent when δ'_M is small. Therefore, a fine discretization is absolutely necessary to demonstrate the effect. Note that W' is about 1 in the central region and about 0.5 along the border, in agreement with eqn 5.21d.

Properties at large densities

W' is averaged over all the fractures of a network. Moreover, the 12-fold symmetry of the hexagonal shape is taken into account and W' is averaged over the triangle $\widehat{OMV}$ as defined in Fig. 5.12a.

The results displayed in this figure are averaged over two independent network realizations. $\langle W' \rangle$ is normalized by W'_c which is the mean value of W' in a disk of radius $R/5$ around a fracture center. The value of $\langle W' \rangle_c$ and the normalized permeability K_{nr}/K_{nSr} are detailed by Mourzenko et al. (2011b), together with the other characteristics of these networks.

The maps in Fig. 5.12 all feature the same general trend. $\langle W' \rangle$ is nearly uniform and equal to $\langle W' \rangle_c$ in the central region. Then, $\langle W' \rangle$ decreases sharply in a peripheral strip of width $\bar{s}$ along the fracture border ∂A. $\bar{s}$ which is defined as the mean spacing between successive intersections along a trace in the fracture, can be shown to be equal to P/ρ' (see eqn 3.14e). A sharper decrease of $\langle W' \rangle$ is observed in the corners of the polygons. In other words, $\langle W' \rangle$ tends towards $1/2$ in a peripheral region whose width is inversely proportional to the fracture density; this

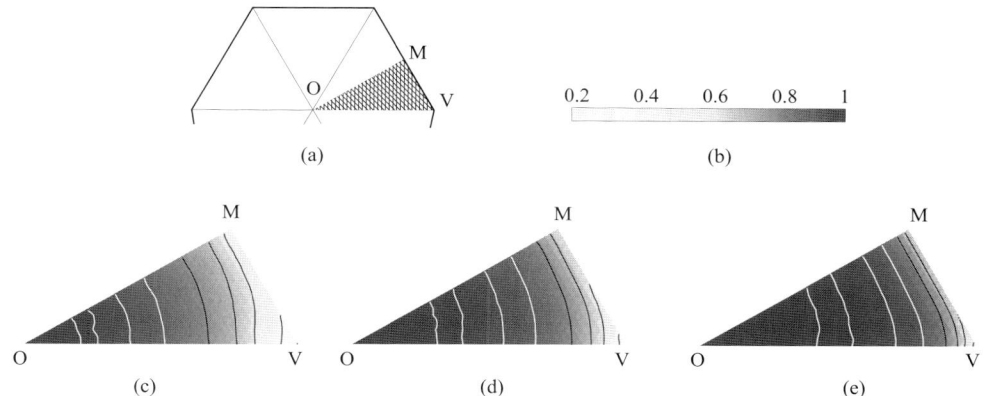

Fig. 5.12 Maps of $\langle W'\rangle/\langle W'\rangle_c$ measured in two independent networks of hexagonal fractures. The triangle $\widehat{OMV}$ is defined in (a) and the color code is shown in (b). Data are for: $\rho' = 10$ (c), 30 (d), 60 (e). Isocontour lines are drawn for the values 0.99, 0.98, 0.95, 0.90 (white) and 0.8, 0.7, 0.6, 0.4 and 0.2 (black), from the center to the border of the fracture.

property corresponds in 2D to the zero fluxes in the segments AB and CD of Fig. 5.10.

A first approximation

This analysis was used by Mourzenko *et al.* (2011b) to devise a simple conceptual model to improve Snow's formula for large densities. The fracture area A can be decomposed into an inner region A_I without contact with the contour ∂A, and a border region A_B along the contour. The typical width of the border region is the mean spacing $\bar{s}$ between intersections provided by eqn 3.14e. Therefore, in a first-order approximation,

$$A_B = P\bar{s} = \frac{P^2}{\rho'}, \quad A_I = A - A_B \qquad (5.22)$$

The idea is to weigh the fracture area in Snow's formula (5.9b) by a value of $\langle W'\rangle$ which depends on the distance of a surface element to its embedding fracture border. The simplest way to do this is to consider that eqns 5.19 and 5.20 apply to the inner region A_I and the border region A_B of each fracture, respectively. Therefore,

$$\mathbf{K} = \sigma \left[1 \times \frac{A_I}{A} + \frac{1}{2} \times \frac{A_B}{A} \right] \int S(\mathbf{n})\,(\mathbf{I} - \mathbf{nn})\,\mathrm{d}\mathbf{n} \qquad (5.23a)$$

which results in the following prediction for the constant in eqn 5.17

$$K_0' = \frac{2PR}{3A} \qquad (5.23b)$$

The corresponding values are shown in Table 5.1 for the various fracture shapes. They are seen to be in very good agreement with the numerical observations.

Exercises 5.2 and 5.3 are recommended to the reader.

5.5 Extensions

This section summarizes several extensions of the results obtained for isotropic and homogeneous networks. The first extension addresses fractures with power-law size distributions (see eqns 5.27–5.29), the second one monodisperse anisotropic networks (5.36), and the third one heterogeneous fractures (5.38).

5.5.1 Fracture networks with power-law size distributions

This subsection is a summary of the paper by Mourzenko et al. (2004). In order to avoid superfluous repetitions, the reader is invited to read Section 3.8.1 where the necessary definitions and notations are presented. Polydisperse networks are illustrated in Fig. 3.8.

The only missing item is fracture transmissivity. σ is considered to be constant for each fracture. Because of the classical Poiseuille law, the typical transmissivity σ_0 of a fracture is expected to be of the order of

$$\sigma_0 = \frac{b^3}{12} \tag{5.24}$$

Fracture lateral sizes and apertures are, in many cases, positively correlated. When both characteristics follow power-law distributions, a scaling relationship can be tentatively written as

$$b = FR^{\kappa_b} \tag{5.25}$$

The existing data in the literature show that the scaling exponent κ_b varies between 0.5 and 2. The original references are provided by Mourzenko et al. (2004). The scaling relationship (5.25) and the cubic law (5.24) imply

$$\sigma' \propto R'^{\beta_\sigma} \tag{5.26}$$

where $\beta_\sigma = 3\kappa_b$, with a possible range of variations $1.5 < \beta_\sigma < 6$.

Systematic calculations were performed in the following conditions. Random networks were generated for cell sizes L larger than or equal to $4R_M$. The exponent a ranges between 1.5 and 2.9 (cf. eqn 3.26); monodisperse networks were also generated. The ratio $R'_m = R_m/R_M$ between the smallest and the largest values of R ranges between 0.01 and 1. The exponent β_σ is chosen between 0 and 6. The dimensionless density ρ'_3 is between 1 and 20. Various shapes and mixtures of shapes were investigated. At least 25 realizations were generated for each set of parameters and permeability was averaged over these realizations.

Results can be summarized by a single formula which is illustrated in Fig. 5.13

$$K_n = \rho \langle \sigma A \rangle K'_2 (\rho'_3) \tag{5.27}$$

This model involves two factors. The extensive term $\rho \langle \sigma A \rangle$ is a measure of network density, weighted by the individual fracture transmissivities. The dimensionless function K'_2 is fairly universal, and the influence

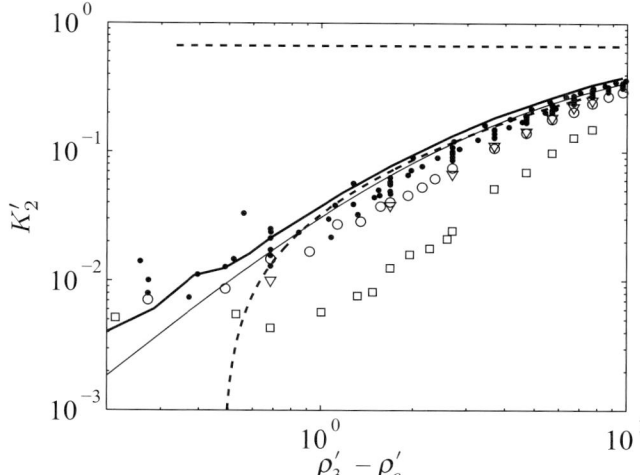

Fig. 5.13 Normalized effective permeability K'_2 for non-periodic networks of regular hexagons versus $\rho'_3 - \rho'_c$ for $\rho'_c = 2.31$. Black dots (●) correspond to the numerical data for $\beta_\sigma = 0$ with $a = 1.5, 2.0, 2.5$ or 2.9, $\beta_\sigma = 1.5$ with $a = 1.5$ or 2.9, and $\beta_\sigma = 3$ with $a = 1.5$. Other symbols are for $\beta_\sigma = 3$ and $a = 2.9$ (∇), $\beta_\sigma = 6$ and $a = 1.5$ (○) or $a = 2.9$ (□). In all cases, $R'_m = 0.1$, and $L/R_M = 4$, except for $a = 1.5$, $\beta_\sigma = 0$ with $L/R_M = 6$. Data for periodic monodisperse fracture networks are given by the thick solid line. The thick broken line corresponds to 2/3 as predicted by the Snow equation (cf. eqn 5.28). The thin broken line is the prediction of eqn 5.29a. The thin solid line is eqn 5.29b with the fitted values for hexagons given in Table 5.1.

of the fracture shape and of the parameters (a, R'_m) of their size distribution are incorporated in the dimensionless density ρ'_3. This relation is valid for all a in a range of $1.5 < a < 3$ and does not vary with R'_m and the fracture shape; it was also conjectured that it is applicable for any value of the exponent a. Note that in terms of K'_2, the Snow equation (5.10) reads as

$$K'_2(\rho'_3) = \frac{2}{3} \quad (5.28)$$

In case of a varying fracture transmissivity σ, eqn 5.27 is valid for $a \leq 3$ and for small or moderate values of the exponent β_σ in the scaling relation (5.26). It breaks down, however, when β_σ increases up to 6, and exponents a which are larger than 3 have not been investigated with varying fracture transmissivities.

K'_2 approaches zero near the percolation threshold, with fluctuations due to finite size effects, whose amplitude depends on a and on the contrast of fracture sizes in the network. For large densities, K'_2 slowly tends towards the value 2/3 predicted by Snow's model (5.28), although it is always smaller than this prediction for finite densities; the same phenomenon was observed for monodisperse networks by Koudina *et al.* (1998). This is partly due to the fact that not all the fractures contribute to the flow, especially for low fracture density, but also due to non-uniformity of the flux distribution among the fractures as well as to the flow interactions between them, since both factors violate the assumptions leading to the Snow equation.

In view of the universality of eqn 5.27, and of its great practical interest, it may be desirable to model it by applying an analytical formula which could be easy to use. Two such models can be proposed.

The first one is inspired by Hestir and Long (1990) as detailed by Mourzenko *et al.* (2004)

$$K'_2 \approx \frac{2}{3}\left[1 - \frac{1}{C_1(\rho'_3 + C_2)}\right], \quad C_1 = 0.10, \quad C_2 = 6.6 \quad (5.29a)$$

The second model is a recent improvement due to Mourzenko *et al.* (2011b)

$$K'_2 = \frac{2}{3} \frac{\beta (\Delta \rho')^2}{\rho' [1 + \beta (\Delta \rho')]} \qquad (5.29b)$$

with β given by eqn 5.16b or by the fitted values provided in Table 5.1.

Equation 5.29a is of historical interest; it works well for large densities, but it fails to describe the region which is close to the percolation threshold. Equation 5.29b represents a considerable improvement compared with the formula proposed by Koudina *et al.* (1998).

Exercise 5.4 is proposed to the reader.

5.5.2 Anisotropic fracture networks

A series of relations was derived for anisotropic networks in Sections 3.8.2 and 3.8.3. Again for the sake of brevity, notations and definitions in these sections are not repeated.

Snow formula for infinite fractures

Mourzenko *et al.* (2011a) wrote the Snow formula (5.9b) for an anisotropic network of infinite fractures as

$$\boldsymbol{K}_{nS} = \mathcal{S} \sigma \langle \boldsymbol{I} - \boldsymbol{nn} \rangle \qquad (5.30)$$

The correction factor $\boldsymbol{\psi_K}$ is a second order tensor defined by

$$\boldsymbol{\psi_K} = \frac{\boldsymbol{K}_{nS}}{K_{nSr}} \qquad (5.31)$$

where K_{nSr} is the Snow permeability of an I^{2}OUD network given by eqn 5.10.

The polar direction $\boldsymbol{p}_F$ of the Fisher distribution (3.37) is again assumed to be parallel to the z-axis of the coordinate system. When expressed in terms of the spherical coordinates (θ, ϕ) (cf. Fig. 3.4), the components of $\boldsymbol{n}$ are $(\sin\theta\cos\phi, \sin\theta\sin\phi, \cos\theta)$. Therefore,

$$\boldsymbol{I} - \boldsymbol{nn} = \begin{pmatrix} 1 - \cos^2\phi\sin^2\theta & -\cos\phi\sin\phi\sin^2\theta & -\cos\phi\sin\theta\cos\theta \\ -\cos\phi\sin\phi\sin^2\theta & 1 - \sin^2\phi\sin^2\theta & -\sin\phi\sin\theta\cos\theta \\ -\cos\phi\sin\theta\cos\theta & -\sin\phi\sin\theta\cos\theta & 1 - \cos^2\theta \end{pmatrix}$$
$$(5.32)$$

Because of the cylindrical symmetry around the z-axis, it is obvious that the off-diagonal components of $\boldsymbol{K}_{nS}$ are equal to zero and that the first two diagonal components are equal. Therefore,

$$\boldsymbol{K}_{nS} = \begin{pmatrix} K_{nS\perp} & 0 & 0 \\ 0 & K_{nS\perp} & 0 \\ 0 & 0 & K_{nS\parallel} \end{pmatrix} \qquad (5.33a)$$

or equivalently,

$$= K_{nSr} \begin{pmatrix} \psi_\perp & 0 & 0 \\ 0 & \psi_\perp & 0 \\ 0 & 0 & \psi_\parallel \end{pmatrix} \qquad (5.33b)$$

where K_{nSr} is given by eqn 5.11c and with

$$K_{nS\perp} = \rho\langle A\rangle \sigma \mathcal{B} \int_0^{2\pi} d\phi \int_0^{\pi/2} d\theta \sin\theta \, \cosh[\kappa\cos\theta](1 - \cos^2\phi \sin^2\theta) \tag{5.33c}$$

$$K_{nS\|} = \rho\langle A\rangle \sigma \mathcal{B} \int_0^{2\pi} d\phi \int_0^{\pi/2} d\theta \sin\theta \, \cosh[\kappa\cos\theta](1 - \cos^2\theta) \tag{5.33d}$$

These two integrals can be evaluated by elementary means by introducing the variable $t = \cos\theta$. The resulting expressions are

$$K_{nS\perp} = \rho\langle A\rangle\sigma\left(1 + \frac{1 - \kappa\coth\kappa}{\kappa^2}\right), \quad K_{nS\|} = \frac{2\rho\langle A\rangle\sigma}{\kappa^2}(\kappa\coth\kappa - 1) \tag{5.34a}$$

Equivalently,

$$\psi_\perp = \frac{3}{2}\left(1 + \frac{1 - \kappa\coth\kappa}{\kappa^2}\right), \quad \psi_\| = \frac{3}{\kappa^2}(\kappa\coth\kappa - 1) \tag{5.34b}$$

When κ tends towards 0, $\coth\kappa$ can be expanded as $\kappa^{-1} + \kappa/3 - \cdots$, and $\psi_\perp$ and $\psi_\|$ tend towards 1.

It should be noted that the volumetric area and network connectivity depend on ρ and ρ', respectively. In view of eqn 3.44, the ratio of these two measures of the density depends on the orientation distribution. Hence, the dimensionless form of eqn 5.33a is

$$\mathbf{K}'_{nS} = \frac{K_{nS}R}{\sigma} = \frac{k_S}{\Phi}\rho'\,\boldsymbol{\psi}_K \tag{5.35}$$

where k_S is still given by eqn 5.11d.

Finite fractures

Numerical permeability calculations were conducted for hexagonal and square fractures ($\kappa = 10$, 50 and 200), for triangles and for rectangles with an aspect ratio of 4 ($\kappa = 10$). The correction for the discretization effects can be applied with the same coefficients as for I^{2}OUD networks. In all cases, the sample size is $L/R = 10$, and averages are taken over at least 10 realizations (50 for squares with $\kappa = 50$). Permeability along the z-axis corresponds to $K'_{n\|}$ and permeabilities along x and y are averaged to obtain $K'_{n\perp}$.

$\Phi K'_{n\perp}/\psi_\perp$ and $\Phi K'_{n\|}/\psi_\|$ are plotted in Fig. 5.14. The data for all fracture shapes, directions and degrees of anisotropy are gathered around a single master curve which is very accurately represented by the model (5.16) with no adjustment. Therefore, it can be concluded that the permeability tensor for these anisotropic fracture networks can be written as

$$\mathbf{K}'_n = \frac{K'_{nr}}{\Phi}\,\boldsymbol{\psi}_K \tag{5.36}$$

where K'_{nr} is the dimensionless permeability of isotropic networks. When $\rho' \to \infty$, $K'_{nr} \to k_S\rho'$, and eqn 5.36 converges toward eqn 5.35 as it should.

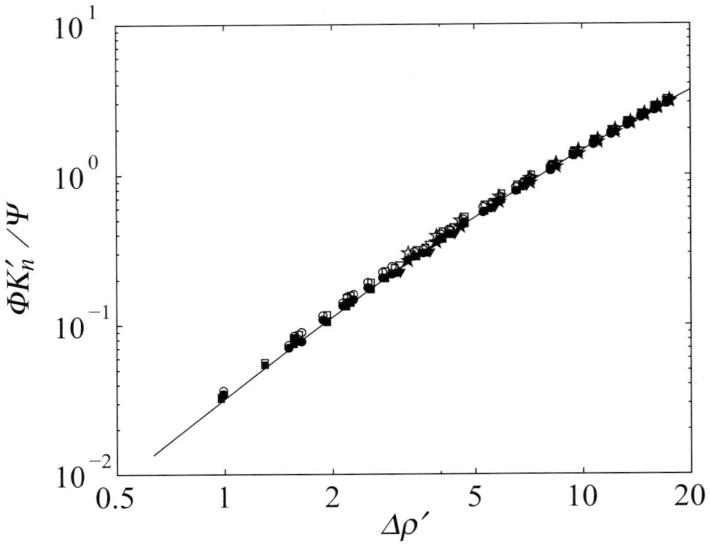

Fig. 5.14 Normalized permeabilities $\Phi K'_n/\psi$ of anisotropic networks as functions of $\Delta\rho' = \rho' - \rho'_{c,r}$. Open and solid symbols correspond to $\Phi K'_{n\parallel}/\psi_\parallel$ and $\Phi K'_{n\perp}/\psi_\perp$, respectively. Data are hexagons (○), squares (□), triangles (∇) and 4-rectangles (⋆), with $\kappa = 10$, 50 and 200. The curve corresponds to eqn 5.16.

5.5.3 Networks of heterogeneous fractures

The definition, the notations and the percolation properties of these networks were presented in Section 3.8.5. Numerical calculations were restricted to hexagonal fractures.

The dimensionless permeability K'_{nh} of a network of heterogeneous fractures is expected to depend on a long list of parameters

$$K'_{nh} = \frac{K_{nh}R}{\sigma} = K'_{nh}(\rho', S_0, \ell_c/R, L/R, \delta_M/R) \qquad (5.37)$$

where S_0 is the fractional open area (2.22c). Only the first three parameters are physically meaningful. The last two are artificial and are introduced for the numerical calculations only; their influence was studied by Hamzehpour et al. (2009) and is skipped here.

An important parameter is the ratio ℓ_c/R which compares the correlation length to the fracture size. The limit $\ell_c/R = 0$ corresponds to site percolation where the transmissivity σ_0 of each triangle is chosen independently of its neighbors. The other limit $\ell_c/R = \infty$ is simpler; each fracture has a uniform transmissivity which is equal either to 0 or to σ_0; therefore, K'_{nh} is expected to depend only on the product $S_0\rho'$ in this limit.

A complete set of data is given in Fig. 5.15. These data which represent a significant amount of numerical calculations have been restricted to an intermediate grid size $\delta_M/R=1/8$. The influence of ℓ_c/R on K'_{nh} is relatively limited and it diminishes when S_0 increases. In the worst case, for $S_0 = 0.4$, K'_{nh} for $\ell_c/R = \infty$ is less than three times greater than for $\ell_c/R = 0$. It should be noticed that the limiting case, $\ell_c/R=0$, is somewhat singular and corresponds to an uncorrelated fracture.

It turns out that the macroscopic permeability K'_{nh} can be rationalized by applying a mean field approximation as introduced by Kirkpatrick

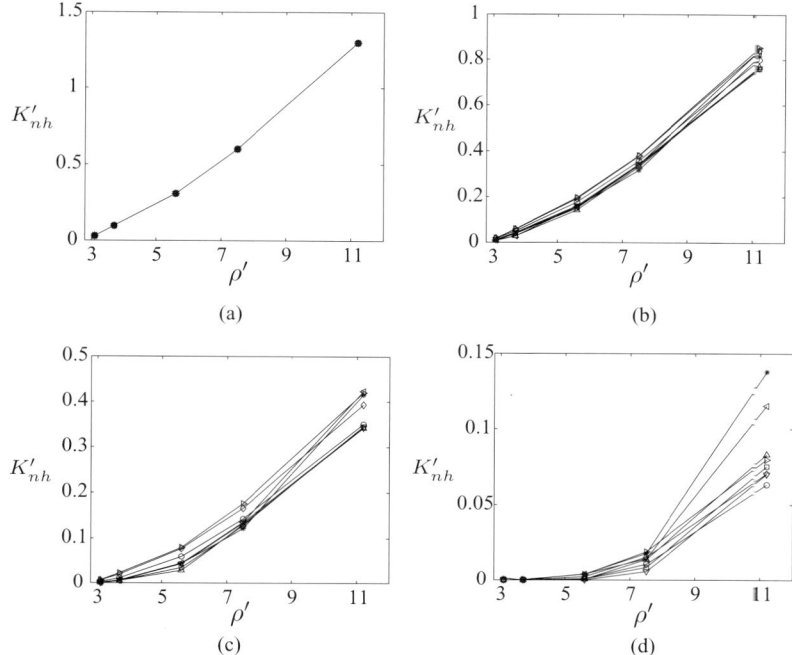

Fig. 5.15 Influence of the ratio ℓ_c/R on the macroscopic permeability K'_{nh}. Data are for $S_0 = 1$ (a), 0.8 (b), 0.6 (c) and 0.4 (d); $\ell_c/R = 0$ (▷), $\frac{1}{8}$ (◊), $\frac{1}{2}$ (○), 1 (▽), $\frac{3}{2}$ (□), 2 (△), 5 (◁), 10 (∗). $\delta/R = \frac{1}{8}$.

(1971). First, calculate the transmissivity $\overline{\sigma}(S_0, \ell_c/R)$ of an infinite fracture made up of a spatially periodic pattern of identical unit cells of size R; the content of the unit cell is characterized by its open surface S_0 and the correlation length ℓ_c; this calculation can be done as described in Chapter 4. The mean field approximation consists in assigning the transmissivity $\overline{\sigma}(S_0, \ell_c/R)$ to each fracture of the network. In other words, one does not take into account the specific open surface of each fracture and its interaction with the others.

Therefore, the mean field approximation of the macroscopic dimensionless permeability K'_{nh} of a network with fractures of transmissivity $\overline{\sigma}(S_0, \ell_c/R)$ is expressed as

$$K'_{nh} = \overline{\sigma}(S_0, \ell_c/R) K'_{nr}, \tag{5.38}$$

where K'_{nr} is the dimensionless permeability of isotropic networks which is discussed in Section 5.4. In Fig. 5.16, this approximation is verified for transmissivity obtained with $\ell_c/R = 0.5$. Therefore, even in this intermediate range of values for ℓ_c/L, a very good agreement with the full numerical results is obtained as soon as S_0 is large enough.

This approximation is exact in the limit $\ell_c/R = \infty$. Data presented by Hamzehpour et al. (2009) show that this also applies to $\ell_c/R = 0$. Since the two limiting cases $\ell_c/R = 0$ and ∞ obey the mean field approximation, agreement between the numerical data and this approximation as a function of $S_0 \rho'$ remains relatively good for all the values of ℓ_c/R.

The success of this mean field approximation is quite remarkable since the binarization of transmissivity creates complex channels which must

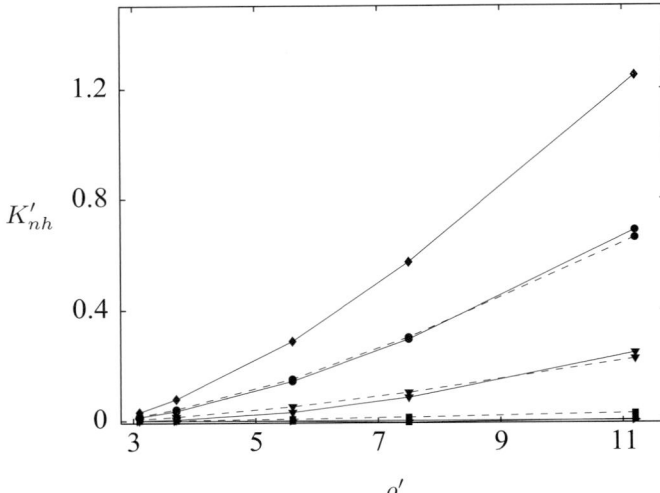

Fig. 5.16 Comparison of the mean field approximation (dashed lines, eqn 5.38) with the numerical results obtained by solving the flow equation in the network (dots). Data are for $S_0 = 1$ (♦), 0.8 (●), 0.6 (▼), 0.4 (■) and 0.2 (▶). $\ell_c/R = 0.5$ and $L/R = 5$.

be present along the same segment at a fracture intersection in order to connect the two fractures.

Therefore, the extension of the classical laws to the present case is a very interesting feature from both theoretical and practical points of view. Moreover, in view of the success of this mean field approximation, it is predicted that a continuous distribution of transmissivities inside the fractures will not yield a significantly different result.

Exercises

(5.1) Show that the operator $\boldsymbol{I} - \boldsymbol{nn}$ projects any vector $\boldsymbol{v}$ onto the plane perpendicular to $\boldsymbol{n}$.

(5.2) Compare the Snow formula and our empirical formula.
When do they give the same results?
Provide some practical guidelines.

(5.3) Consider the three networks studied in Exercise 3.7. Namely: $R = 2.5$m, $\rho = 1.5$, 4.2 and 7.7. Each fracture is a plane channel $b_m = 10^{-4}$ m as in Exercise 4.1. Therefore, $\sigma = 8.33 \cdot 10^{-14}$ m^3.

  (i) Determine the permeability of these three networks by the Snow formula and formula 5.16.

  (ii) Determine the flow rate through a fractured block $1 \times 1 \times 1$ m^3, $R = 2.5$m, of dimensionless permeability $K'_{nr} = 1$ submitted to a pressure gradient of 10^3 Pa/m. Compare it with the flow rate in a plane channel (see Exercise 4.1).

(5.4) Consider networks of polydisperse fractures characterized by an exponent a. The fractures are triangles, hexagons and 6-rectangles. They are inscribed in circles whose minimal and maximal radii are R_m and R_M, respectively. The values of these quantities are the same as in Exercise 3.9; $a = 2$, $R_m = 0.1$ m, and $R_M = 1$ m.

  (i) Determine the critical dimensionless density ρ'_{3c} for these three shapes.

  (ii) Calculate the dimensionless permeability K'_{nr} for $\rho'_3 = 2\rho'_{3c}$, $4\rho'_{3c}$ and $10\rho'_{3c}$ by eqn 5.29b with β provided by eqn 5.16b. Deduce the dimensional permeability K_{nr}.

  (iii) Redo question (ii) with β provided by Table 5.1.

  (iv) Comment on the results.

Transport in a fractured porous medium

6

6.1	Introduction	109
6.2	General	109
6.3	Numerical methodology	113
6.4	Fractured porous media with I²OUD fractures	118
6.5	Extensions	121
Exercises		128

6.1 Introduction

A *fractured porous medium* is composed of a porous medium with non-zero permeability and fractures. When a fluid flows through a fractured porous medium, it flows through the porous medium and through the fractures with transfers between these two structures. Each point in the porous medium can be assigned a bulk permeability K_m while, as in Chapter 5, each point in the fractures can be assigned a surface transmissivity σ. It should be remembered that K_m and σ are homogeneous to the square and to the cube of a length, respectively.

This chapter, which is mostly based on two papers by Bogdanov *et al.* (2003a) and Bogdanov *et al.* (2007), is organized in the following way. Section 6.2 is devoted to a general presentation of the equations, to the determination of permeability and to a simple approximation. Section 6.3 very quickly spans the general methodology which is used to mesh the porous medium, to discretize the equations and to solve them; actually, the most difficult part is the first one and it is a key point in this field. An elementary example based on the network used in Subsection 5.3.4 is provided in Subsection 6.3.4. I²OUD networks of identical fractures which are isotropically and uniformly distributed are addressed in Section 6.4; several effects are discussed and master curves which are valid for fractures of any shape are presented.

Extensions of the results to fractured porous media with power-law size distribution are presented and discussed in Section 6.5, where slightly compressible flows with application to pressure drawdown well tests are also summarized.

6.2 General

6.2.1 Flow equations

The physical situation is sketched in two dimensions in Fig. 6.1. The drawing is in 2D since it is easier to do than in 3D, but everything is 3D of course. As a rule, we are not interested in 2D calculations for the simple reason that, whether you like it or not, the world is in 3D!

Let the porous rock matrix have a bulk permeability K_m $[L^2]$ that can vary with space. The local seepage velocity $\overline{v}$ is provided by Darcy's law

110 *Transport in a fractured porous medium*

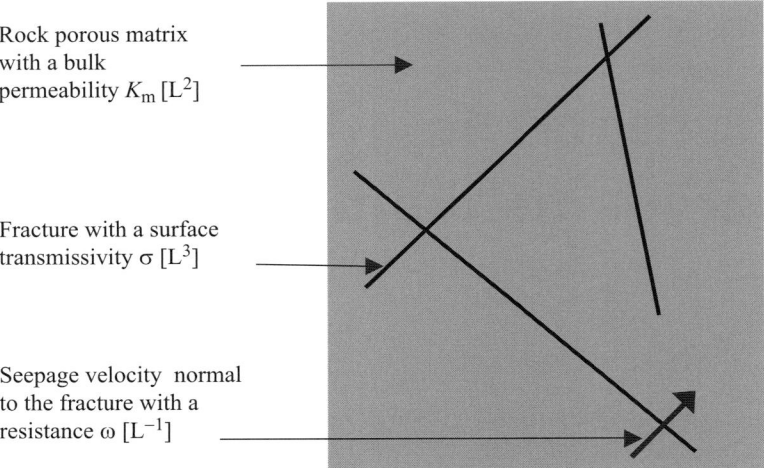

Fig. 6.1 2D sketch of a 2D fractured porous medium.

$$\overline{\boldsymbol{v}} = -\frac{K_m}{\mu} \nabla p \qquad (6.1)$$

where μ is fluid viscosity and p is pressure. The overbar over ∇p is omitted for the sake of simplicity. The continuity equation for the local seepage velocity in the porous matrix reads as

$$\nabla \cdot \overline{\boldsymbol{v}} = 0 \qquad (6.2)$$

We assume that the hydraulic properties of a fracture can be characterized by two effective coefficients, namely a fracture transmissivity σ $[L^3]$ and a cross resistance ω $[L^{-1}]$. They can be defined by considering two situations where the main flow direction is set parallel and normal to the fracture plane, respectively. In the first case, the flow rate $\boldsymbol{J}$ is usually defined per unit width of the fracture; $\boldsymbol{J}$ is related to the surface pressure gradient $\nabla_s p$ by the two-dimensional Darcy's law

$$\boldsymbol{J} = -\frac{\sigma}{\mu} \nabla_s p \qquad (6.3)$$

Hence, it is clear that the fracture transmissivity σ is homogeneous to the cube of a length because the flux $\boldsymbol{J}$ is taken per unit width of the fracture. This can be compared to eqn 6.1 where $\overline{\boldsymbol{v}}$ is a flux per unit area. In the second case of a flow which is normal to the fracture plane, the seepage velocity $\overline{\boldsymbol{v}}_\perp$ normal to the fracture induces a pressure drop Δp expressed by

$$\overline{\boldsymbol{v}}_\perp = -\frac{1}{\mu\omega} \Delta p \qquad (6.4)$$

where ω is the normal resistance of the fracture.

Many different situations can be described by this simple formalism. First suppose that the fractures are empty and that they can be described by an equivalent aperture b_S (cf. eqn 4.17); then,

$$\sigma = \frac{b_S^3}{12} \qquad \omega = 0 \qquad (6.5a)$$

This value of ω implies that flow across the fracture does not induce any pressure drop. Second, suppose that the fracture has been filled with an impermeable material (such as shales or veins); then,

$$\sigma = 0 \qquad \omega = \infty \qquad (6.5b)$$

Third, consider a fracture which is empty, but whose walls have been partially clogged by a chemical,

$$\sigma = \frac{b_S^3}{12} \qquad \omega \neq 0 \qquad (6.5c)$$

The last typical case is that of a plane channel of aperture b, filled with a porous material with permeability K_f. Knowledge of K_f and b is sufficient to determine σ and ω

$$\sigma = bK_f \qquad \omega = \frac{b}{K_f} \qquad (6.6a)$$

and inversely,

$$K_f = \sqrt{\frac{\sigma}{\omega}} \qquad b = \sqrt{\sigma\omega} \qquad (6.6b)$$

Hence, the two independent variables σ and ω span many different situations. σ and ω can be expressed in dimensionless form as

$$\sigma' = \frac{\sigma}{R_o K_m}, \quad \omega' = \frac{K_m}{R_o}\omega \qquad (6.6c)$$

where R_o is a characteristic value of half the lateral extension of the fractures. In the rest of this chapter, we shall mostly address the last case summarized by these relations (6.6).

The mass conservation equation for the flow in a fracture reads

$$\nabla_s \cdot \boldsymbol{J} = -(\overline{\boldsymbol{v}}^+ - \overline{\boldsymbol{v}}^-) \cdot \boldsymbol{n} \qquad (6.7)$$

where $\boldsymbol{n}$ is the unit vector normal to the fracture plane which can be oriented in two equivalent ways, $\overline{\boldsymbol{v}}^+$ is the seepage velocity in the matrix on the side of $\boldsymbol{n}$ and $\overline{\boldsymbol{v}}^-$ is the seepage velocity on the opposite side.

The transport equations (6.1–6.4, 6.7) have to be supplemented with macroscopic boundary conditions that are going to be presented and justified in the next subsection.

6.2.2 Permeability K_{eff} of a fractured porous medium

For the sake of completeness, the basic flow experiment to determine permeability K_{eff} of the fractured porous medium is recalled in Fig. 6.2. It is very similar to Figs 5.2b and 1.4. It should be noticed that the experimentalist does not need to know whether the porous medium is permeable or not to perform the experiment. Again, the fractured porous

112 *Transport in a fractured porous medium*

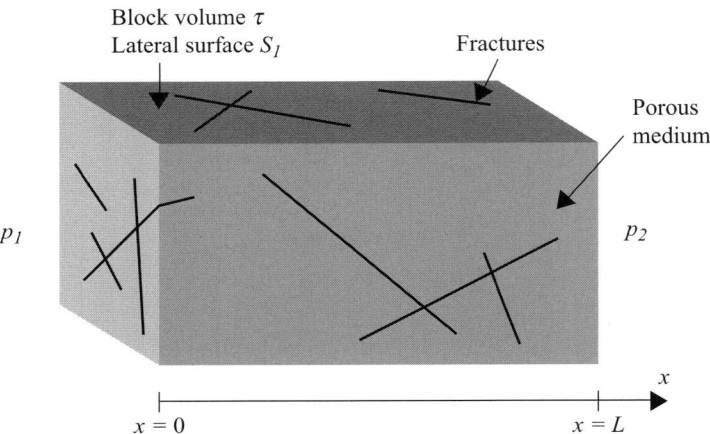

Fig. 6.2 Determination of the permeability K_{eff} of a fractured porous medium.

medium is wrapped in an impermeable cover and set between two vessels with two pressures p_1 and p_2.

Therefore, one has to add the two overall boundary conditions to the previous equations

$$p(x=0) = p_1, \quad p(x=L) = p_2; \quad \boldsymbol{J} \cdot \boldsymbol{n} = 0 \text{ and } \overline{\boldsymbol{v}} = 0 \text{ on } S_l \quad (6.8)$$

The overall seepage velocity $\overline{\overline{\boldsymbol{v}}}$ is obtained by integrating the flow rates over the fracture surfaces and the porous medium, and by dividing the result by the total volume τ of the block (cf. Fig. 6.2). Since the equations are linear, every quantity is proportional to the overall pressure gradient $\overline{\overline{\frac{\partial p}{\partial x}}}$. Therefore, $\overline{\overline{\boldsymbol{v}}}$ can be expressed as (cf. Adler and Thovert, 1999)

$$\overline{\overline{v}}_x = \frac{1}{\tau} \left\{ \int_{\tau_m} \overline{v}_x \mathrm{d}\tau + \int_{S_f} J_x \mathrm{d}s \right\} = -\frac{K_{\text{eff}}}{\mu} \overline{\overline{\frac{\partial p}{\partial x}}} \quad (6.9)$$

where τ_m is the matrix volume, S_f the surface of all the fractures and $\overline{\overline{\frac{\partial p}{\partial x}}} = (p_2 - p_1)/L$. In the general case of an anisotropic fractured porous medium, K_{eff} is a tensor and the previous equation is made tensorial, but, such an extension is not necessary for the rest of this chapter.

A dimensionless permeability K'_{eff} can be defined as

$$K'_{\text{eff}} = \frac{K_{\text{eff}}}{K_m} \quad (6.10)$$

It should be noted that the permeability units are not identical for the two dimensionless permeabilities K'_n (cf. eqn 5.6) and K'_{eff}.

6.2.3 A simple approximation for K_{eff}

The simplest approximation which can be made is to assume that K_{eff} is the sum of the permeability of the porous medium and the permeability K_n of the fracture network

$$K_\text{eff} = K_m + K_n \qquad (6.11\text{a})$$

or in dimensionless terms (cf. eqn 6.6c)

$$K'_\text{eff} = 1 + \sigma' K'_n \qquad (6.11\text{b})$$

σ' is an important dimensionless parameter which compares the transmissivity of the fracture with the permeability of the porous medium. Network permeability can be further approximated by eqn 5.11c derived for an isotropic network of infinite fractures with the same surface area $\mathcal{S}$ per unit volume as the real network. Therefore, with a subscript S to recall the Snow equation (5.11c)

$$K_{\text{eff}S} = K_m + \frac{2}{3}\sigma\mathcal{S} = K_m + \frac{4}{3}\sigma\frac{\rho'}{P} \qquad (6.12\text{a})$$

where P is recalled as the perimeter of the fractures. Equivalently,

$$K'_{\text{eff}S} = 1 + \frac{4}{3}\frac{\sigma\rho'}{PK_m} = 1 + \frac{4}{3}\frac{\sigma'\rho'}{P'} \quad \text{with} \quad P' = \frac{P}{R} \qquad (6.12\text{b})$$

For circular disks, $P = 2\pi R$

$$K'_{\text{eff}S} = 1 + \frac{2}{3\pi}\sigma'\rho' \qquad (6.12\text{c})$$

This section is illustrated in the Exercise 6.1.

6.3 Numerical methodology

The content of this section, which could be very technical and somewhat tedious, is not detailed. Just as in Section 5.3, the most difficult point is the meshing of the porous space located in between the fractures. Indeed, the discretization and the numerical resolution of the local equations are much easier.

6.3.1 Meshing

The situation is in a way the same as for the meshing of fractures in Section 5.3, but with dimensions which are larger than before by 1. In fractures, one starts from segments of size δ with a maximal value equal to δ_M (see Section 5.3.1) and one completes the fracture surface by adding triangles whose sides are about δ, until the surface is entirely covered. Here, one starts from the triangles defined in the fractures and one completes the volume in between by adding tetrahedra whose sides are about δ, until the volume is entirely covered. The method which has been chosen is the advancing front method just as for fractures.

The fractures provide the boundary surface enclosing the 3D domain to be covered by the mesh, i.e. the initial front of the generation process. It is represented by the list of the oriented triangular faces of the 2D triangulation of the fracture network.

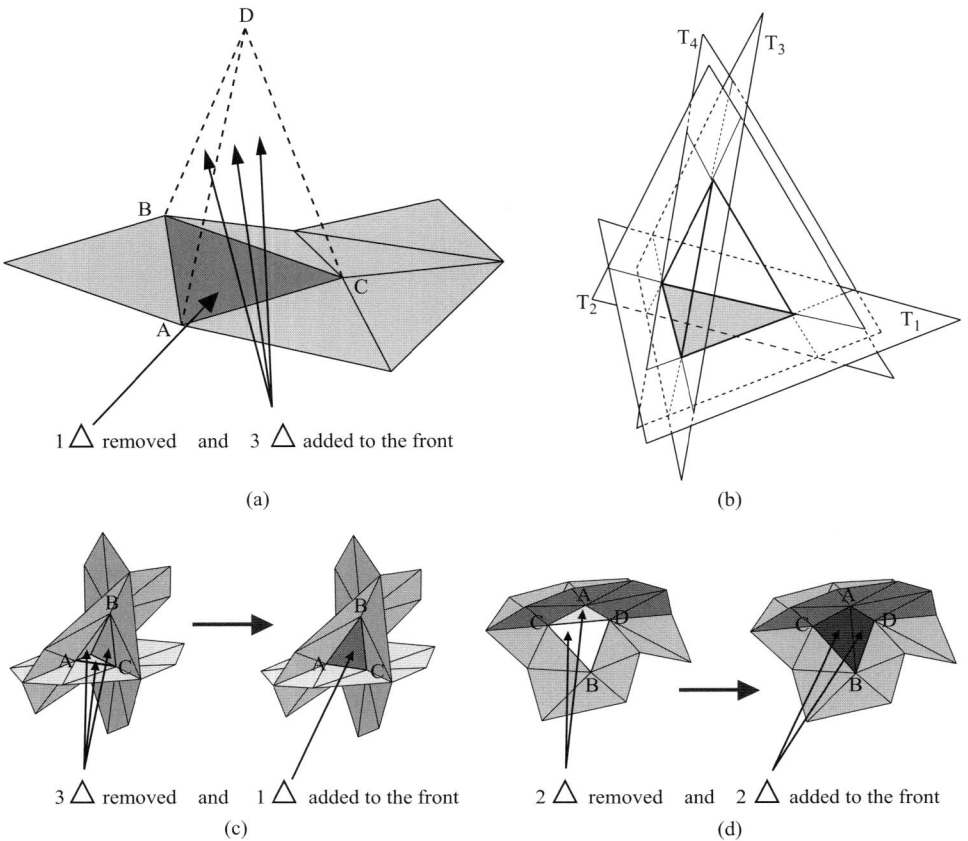

Fig. 6.3 Meshing of the porous space located in between the fractures. (a) Insertion of a new point. (b) Existing tetrahedron between triangles. (c) Insertion of the triangle ABC. (d) Insertion of the two triangles ABC and ABD.

Then space is progressively covered by tetrahedra. The basic step consists of adding a fourth point to an existing triangle in the front in order to build a tetrahedron (see point D in Fig. 6.3a). The new point is inserted at an equal distance δ from the three vertices of a triangle in the front, if no other grid point is closer than δ. Initially, δ is equal to the maximum segment length δ_M in the 2D mesh of the fracture network, but this value is slightly modified as the process goes on.

Whenever possible, tetrahedra are formed by connecting triangles at the front without inserting any new point. This process is illustrated in Figs 6.3b, c and d. In (b), a tetrahedron is already defined between three existing triangles; in (c), there is a sort of cavity; it suffices to close it with an extra triangle ABC to define a new tetrahedron. In (d), there is a kind of passage between two triangles; if points A and B are not too far apart, one can close the corresponding space with two triangles ABC and ABD.

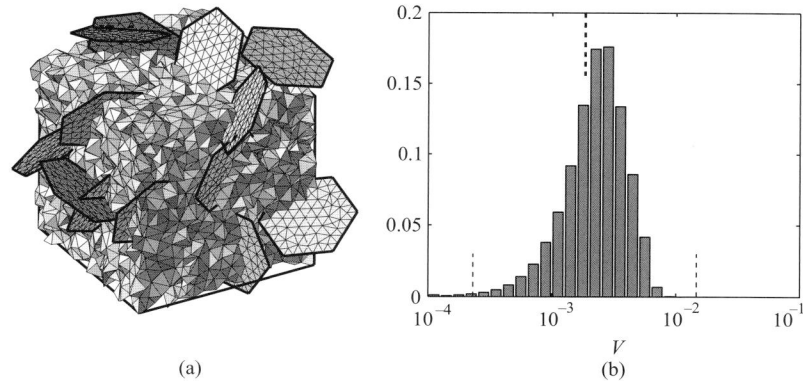

Fig. 6.4 Meshing of the medium located in between the fractures. (a) Example of meshing of the whole space by tetrahedra which locally coincide with the triangles in the fractures. (b) Volumetric histogram of the volumes of the tetrahedra in the previous meshed porous medium; dashed lines: volumes of regular tetrahedra with edges $\delta_M/2$, δ_M, $2\delta_M$.

Hence, the front is progressively deformed by insertion of new faces, when a new point is inserted, and deletion of the existing triangles which are used to form tetrahedra. The number of faces in the front varies and the process terminates when it is reduced to zero. As the algorithm proceeds, one can see that the meshed portion of space makes progress, hence the name advancing front technique.

Three-dimensional random fracture networks may have a very complex topology. For example, several disconnected fracture clusters can exist for small densities, whereas the matrix can be subdivided by the fractures into separate blocks for large densities. Automated procedures need to be implemented in order to cope with these situations.

At the end of the process, the unmeshed domain is reduced to a number of small, but irregular polyhedra. Local mesh refinement is applied in order to resolve situations which cannot be triangulated with a standard technique, such as Schönhart polyhedra.

Actually, this technique is much more complex than it might seem at first sight; in particular, round-off errors play an unexpected and important role, especially in regions where many fractures mutually intersect.

A last remark should be made. One key factor for progress is to decrease δ. But, here as well, one should find a good compromise. If δ does not decrease fast enough, no progress is made. If δ decreases too fast, the volume is populated with a large number of very small tetrahedra without making any significant progress.

The meshing and its efficiency is a key factor in the precise resolution of flow equations. We shall return to this point in Chapter 8.

The meshing process can be illustrated by a couple of examples. A meshed fractured porous medium is shown in Fig. 6.4a. A mesh can be considered to be good when it enables precise and rapid numerical calculations; the discretized equations should not contain coefficients which are too close to zero. A necessary condition is that the volumes of the tetrahedra should not be too small; this happens under two circumstances: when a tetrahedron has all its sides small or when it is almost flat.

Therefore, a first estimation of the quality of the mesh is provided by the volumetric histogram of the tetrahedra; such a histogram is shown in Fig. 6.4b. Three reference volumes are indicated, which correspond to regular tetrahedra with edges $\delta_M/2$, δ_M, $2\delta_M$; Fig. 6.4b shows that, with a few exceptions of the very small ones, most created tetrahedra have volumes V in between the two limits

$$\frac{V_M}{8} \leq V < 8V_M \quad \text{with} \quad V_M = \frac{\delta_M^3}{12}\sqrt{2} \qquad (6.13)$$

where V_M is the volume of a regular tetrahedron of side δ_M.

Of course, other meshes can be found in the literature such as the one described by Paluszny et al. (2007).

6.3.2 Discretization of the equations and resolution

At this stage, the rock matrix is discretized by tetrahedral volume elements, with given permeabilities K_m, and the fractures by triangular surface elements of given transmissivities σ. These values can vary with position. The pressure is evaluated at the vertices of these elements.

When the fractures are viewed as vanishingly thin, empty or very permeable layers ($\omega \to 0$), there is no pressure jump between two points facing each other on the two opposite sides of a fracture (cf. eqn 6.4). A single value of the pressure can be used in the numerical formulation per vertex of fracture element. Thus, the mass balance in the single control volume in Fig. 6.5a is implemented to obtain the equation for the pressure at a node.

However, if the fractures are filled with a low-permeability material or if their walls are partially clogged ($\omega \gtrsim b/K_m$), they are obstacles to the flow, and a pressure difference can appear between their two faces. Then, it is necessary to solve for these two values $p^\pm$ of the pressure. For convenience sake, a third value of the pressure p_f is introduced, in the middle plane of the fracture. Three balance equations are written, over the three control volumes displayed in Fig. 6.5b, to obtain the necessary equations. The normal fluxes $\overline{v}^\pm \cdot n$ in eqn 6.7 result from eqn 6.1 in the matrix; these fluxes are related to the pressure drop between the matrix and the fractures by

$$\overline{v}^\pm \cdot n = \pm \frac{2}{\mu\omega}\left(p_f - p^\pm\right). \qquad (6.14)$$

More complex situations occur along the intersection line of two fractures (with five values of p) and at the intersection point of three fractures (with nine values of p).

The balance equations for control volumes around each mesh point, together with pressure drop conditions across the unit cell in the direction of the mean pressure gradient $\overline{\overline{\nabla p}}$, provide a set of linear equations for the pressures, which is solved by using a conjugate gradient algorithm as in Section 5.3.2. The convergence criterion is basically the same. This solution is actually much less demanding in terms of computational resources than the mesh generation.

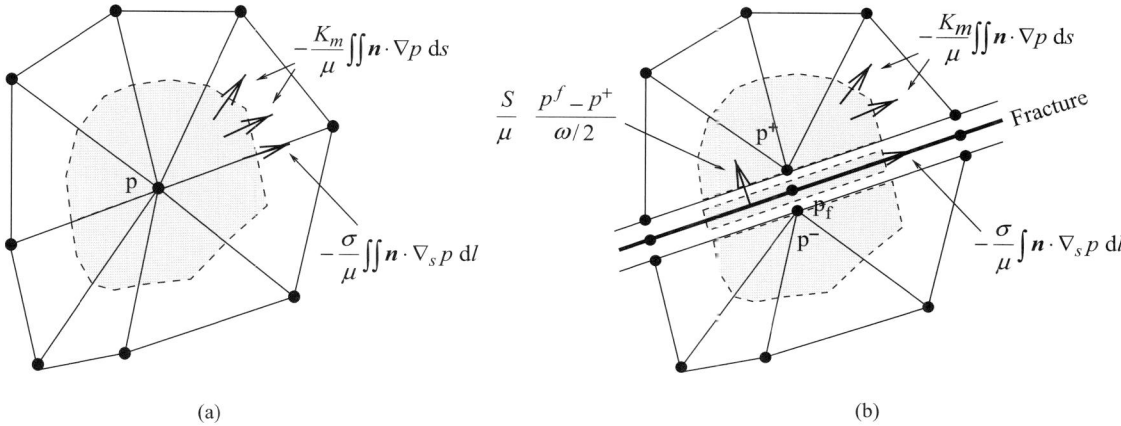

Fig. 6.5 Single (a) and triple (b) control volumes for the mass balance equations at a point on a fracture.

A hybrid finite element–finite volume approach was used by Paluszny et al. (2007).

6.3.3 The permeability of the fractured porous medium

The permeability of the fractured porous medium is derived from the pressure field determined in Subsection 6.3.2. Seepage velocity is obtained by integration over the triangles of all the fractures and over all the tetrahedra of the porous media, according to the first equality in eqn 6.9. Then, K_{eff} is derived according to the second equality in eqn 6.9.

A completely different approach was proposed by Sahimi (2000).

6.3.4 An elementary example

An elementary example can illustrate the effect of the parameter c' on the results. The fracture network is the one used in Section 5.3.4 the fractures do not exert any resistance to cross flow, i.e. $\omega = 0$. The same fracture 4 which intersects four other fractures is shown.

When σ' is equal to 10^4, as in Fig. 6.6a, the fractures are much more conductive than the porous medium (cf. Exercise 6.1) and in the fractures of the fractured porous medium the flow field is expected to be the same as the flow in the fractures of the fracture network. Indeed, a comparison between Figs 6.6a and 5.6 shows that this is the case. This test is very important for a cross verification of the numerical codes which can always be wrong!

The reverse occurs when σ' is small, which means that the fractures are not very conductive. In such a case, the flow field should be identical to the one in a homogeneous porous medium, i.e. it should be constant. This is indeed what is seen in Fig. 6.6d where the flow field is almost constant.

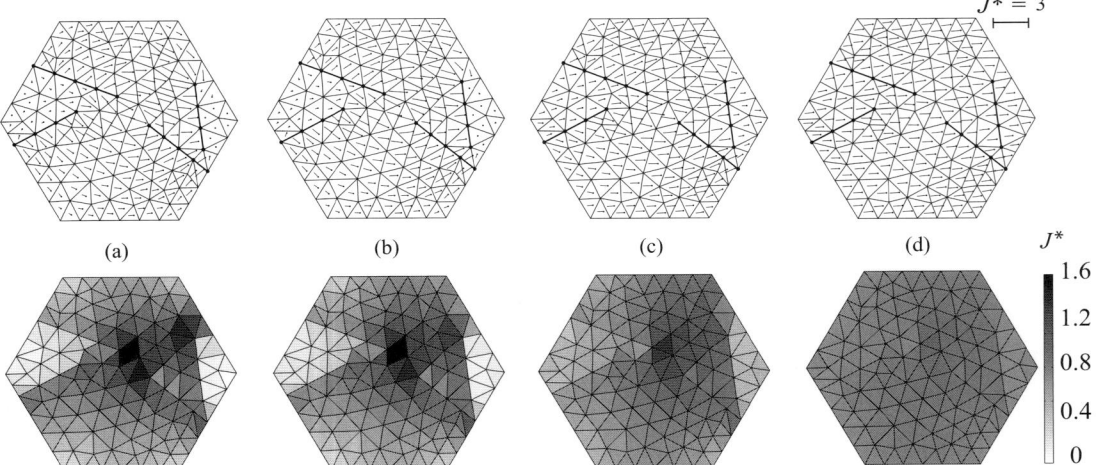

Fig. 6.6 The dimensionless flow rate $J^* = \mu J/\sigma \nabla p$ in fracture 4 of the fractured porous medium whose fracture network is displayed in Fig. 5.6. $\omega' = 0$. Data are for: $\sigma' = 10\,000$ (a), 10 (b), 1 (c), 0.1 (d). The scales for the segments and for the color code are provided at the right.

In between these two extreme cases, the flow fields for $\sigma' = 10$ and 1 are neither identical to the one in a fracture network, nor to the one in a homogeneous porous medium.

6.4 Fractured porous media with I²OUD fractures

Systematic calculations were performed for I²OUD fractures embedded in a porous medium; the corresponding permeability is denoted by $K_{\text{eff},r}$.

6.4.1 Importance of the existence of the percolation threshold on individual samples

Let us start with hexagonal fractures and look at some of the combined effects of the dimensionless density ρ', the dimensionless fracture transmissivity σ' and finite size effects.

A series of results are shown in Fig. 6.7 and they provide some interesting features. The dimensionless aperture b' is defined by

$$b' = \frac{b}{R} \qquad (6.15)$$

Look first at the two extremes. In (a), ρ' is equal to 0.487 which is way below the percolation threshold which is equal to 2.3. As a result, the fracture networks never percolate and it is interesting to see that the macroscopic permeability $K'_{\text{eff},r}$ is hardly influenced by σ'. The fracture

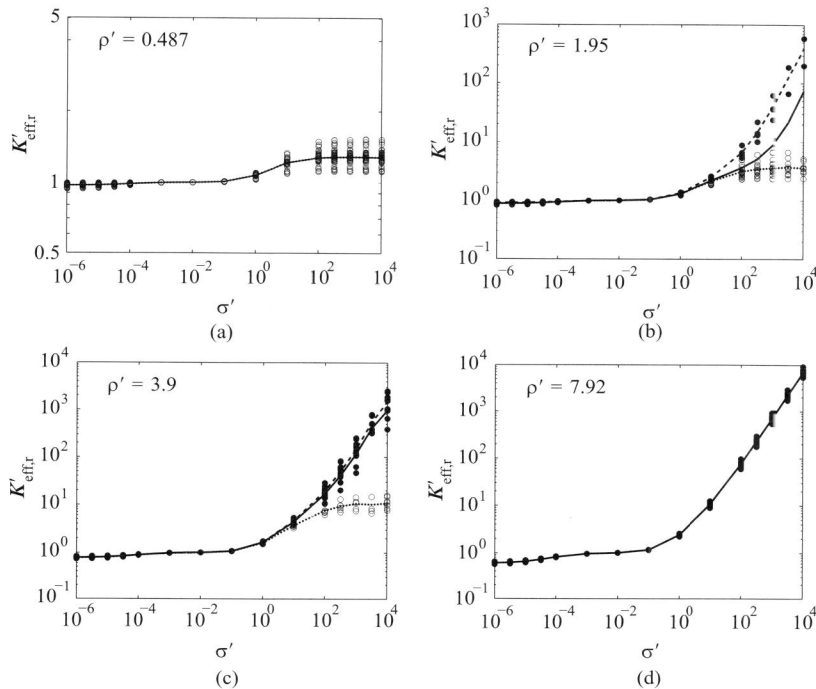

Fig. 6.7 Permeabilities $K'_{\text{eff},r}$ of individual samples, with 4 to 65 hexagonal fractures with aperture $b' = 0.01$ in a unit cell with size $L = 4R$, versus the fracture conductivity σ'. The solid lines are the overall statistical averages $\langle K'_{\text{eff},r} \rangle$. The broken ($----$) and dotted ($\cdots\cdots$) lines are the averages $\langle K'_{\text{eff},r} \rangle_p$ and $\langle K'_{\text{eff}} \rangle_{np}$ over the configurations containing a percolating or a non-percolating fracture network, respectively. Solid and open symbols correspond to percolating and non-percolating networks, respectively.

transmissivity σ' varies by more than eight orders of magnitude while $K'_{\text{eff},r}$ increases by less than 30% on average.

In (d), ρ' is equal to 7.92 which is well above 2.3 and all the networks percolate. Now, σ' has a strong influence on $K'_{\text{eff},r}$. The curve is composed of two parts, depending on whether the matrix or the fractures are dominant. For large values of σ', $K'_{\text{eff},r}$ varies almost linearly with σ' in agreement with the simple approximation (6.12b).

In between these two extremes, one can clearly see the influence of finite size effects. Actually, the data are located on two branches. The lower branch corresponds to the non-percolating networks and the upper one to the percolating ones. Of course, from graph to graph, when density increases, the lower branch disappears progressively to the benefit of the upper one.

There is a very important practical lesson to remember from this figure. When the fractures do not percolate, their influence on macroscopic permeability is small even if they are very conductive. Of course, this influence increases when ρ' increases as can be seen on the lower branch of Figs 6.7a–c.

The same data can be viewed differently by exchanging ρ' and σ' in Fig. 6.8. Again, it is a good policy to look first at the two extremes. For low values of σ' in (a), $K'_{\text{eff},r}$ is a decreasing function of ρ'. This phenomenon is due to the role of b'. In view of eqns 6.6b–c and with $b' = 10^{-2}$ in these calculations

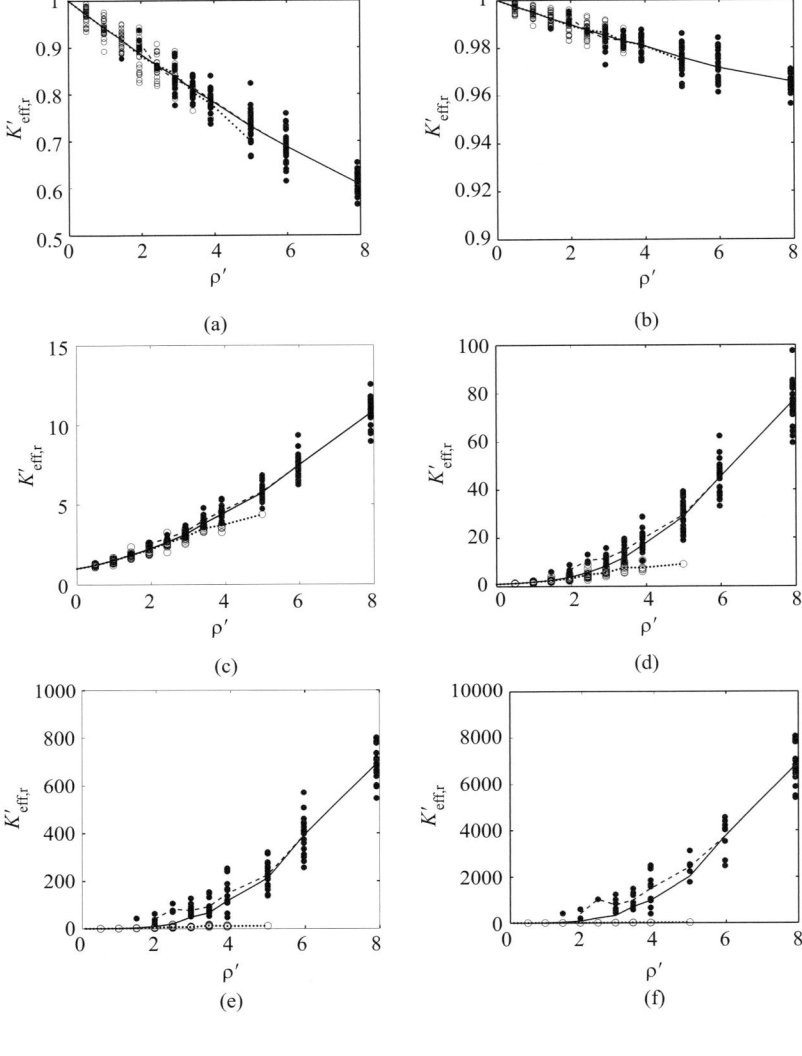

Fig. 6.8 Permeabilities $K'_{\text{eff},r}$ of individual samples, with 4 to 65 hexagonal fractures with aperture $b' = 0.01$ and conductivity σ' in a unit cell with size $L = 4R$, versus the network density ρ'. Data are for: $\sigma' = 10^{-6}$ (a), 10^{-3} (b), 10 (c), 100 (d), 10^3 (e), 10^4 (f). $\omega' = 100$ and 0.1 when $\sigma' = 10^{-6}$ and 10^{-3}, respectively; ω' is negligible in all the other cases. The solid lines are the overall statistical averages $\langle K'_{\text{eff},r} \rangle$. The broken ($---$) and dotted ($\cdots$) lines are the averages $\langle K'_{\text{eff},r} \rangle_p$ and $\langle K'_{\text{eff},r} \rangle_{np}$ over the configurations containing a percolating or a non-percolating fracture network, respectively. Solid and open symbols correspond to percolating and non-percolating networks, respectively.

$$\omega' = \frac{b'^2}{\sigma'} = \frac{10^{-4}}{\sigma'} \qquad (6.16)$$

Hence, the fracture cross resistance ω' is significant when $\sigma' \leq 10^{-3}$ while their transmissivity is negligible. In this case, the fractures play the role of obstacles, and of course, the more numerous the obstacles, the less permeable the medium as a whole.

For large values of σ' in (e), which implies negligible ω', the behavior is totally different and $K'_{\text{eff},r}$ is an increasing function of ρ'; for large densities, the linear variation predicted by the simple approximation (6.12b) is in fact observed.

In an intermediate range with $10^{-3} \leq \sigma'$, $\omega' \leq 10^{-1}$, both transmissivity and cross resistance are small and the fractures are nearly neutral. This corresponds to the central plateaus in Fig. 6.7.

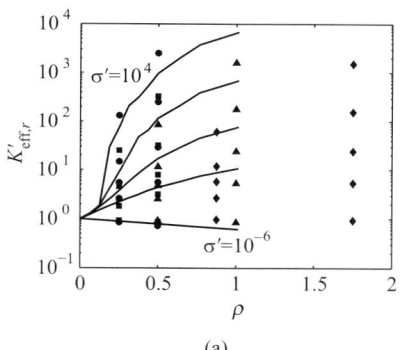 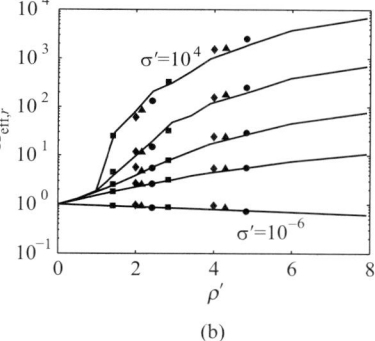

Fig. 6.9 The permeability $K'_{\text{eff},r}$ of fractured porous media as a function of the density and of the fracture transmissivity σ'. (a) Dimensional density ρ. (b) Dimensionless density ρ'. Data are for: squares (■), rectangles with aspect ratios 2 (▲) and 4 (♦), hexagons (solid lines) or 20-gonal (•) fractures. Values of σ' from bottom to top: 10^{-6}, 10, 10^2, 10^3, 10^4.

6.4.2 Influence of the fracture shapes

The influence of the fracture shape is displayed in Fig. 6.9. In (a), $K'_{\text{eff},r}$ is represented as a function of the dimensional density ρ and there is apparently no logic except obvious facts such as the fact that $K'_{\text{eff},r}$ increases with density when σ' is large enough.

But everything reverts back when the same data are plotted as functions of ρ'. Therefore, as in the previous cases, the use of the dimensionless density ρ' enables us to almost totally forget the role of the fracture shape. This is again a very powerful argument in favor of ρ'.

Practically speaking, these curves may serve as master curves to obtain a quick and easy estimate of the macroscopic permeability of a fractured porous medium when ρ' and σ' are known. In other words,

$$K'_{\text{eff},r} = K'_{\text{eff},r}(\rho', \sigma') \tag{6.17}$$

Explicit analytical expressions of this equation are provided in Section 6.5.1 for polydisperse networks. Of course, these expressions also apply to monodisperse networks.

Applications of these master curves are proposed in Exercise 6.2.

6.5 Extensions

The major direct extension which has been made of the previous results is the one for fractured porous media with fractures possessing a power law size distribution with a dimensionless transmissivity given by eqn 5.26.

Another extension in a very different direction is the preliminary but general study of slightly compressible flows and its application to well tests in Section 6.5.2.

6.5.1 Fractured porous media with power-law distribution of fracture sizes

Systematic calculations were performed exactly in the same conditions as for the fracture networks in Section 5.5.1 with $\omega = 0$, i.e. fractures do not oppose any resistance to normal flow. Let us recall these conditions for the sake of simplicity. IOUD random networks are generated for cell sizes L which are larger or equal to $4R_M$. The exponent a ranges between 1.5 and 2.9; monodisperse networks are also generated. The ratio R_m/R_M between the smallest and the largest values of R ranges between 0.01 and 1. The exponent β_σ is chosen between 0 and 6. The dimensionless density ranges between 1 and 20. Various shapes and mixtures of shapes were investigated. At least 25 realizations were generated for each set of parameters, and permeability was averaged over these realizations. The notations provided in Section 5.5.1 are not recalled here.

The macroscopic permeability K_{eff} is expected to be a function of a long list of parameters

$$K_{\text{eff}}, K_n = f(a, R_m, R_M, \rho, \text{shape}, K_m, \sigma_0, \beta_\sigma, L, \delta_M) \qquad (6.18)$$

In this list, there are two artificial parameters which are the size L of the unit cell and the discretization length δ_M. Most of the results presented below are extrapolated to the limits $L \to \infty$ and $\delta_M \to 0$. The dimensionless form of the results is again almost independent of the shape of the fractures.

By analogy with eqn 6.6c, the dimensionless fracture transmissivity is defined as

$$\sigma' = \frac{\sigma}{R_M K_m} \qquad (6.19)$$

An example of results is given in Fig. 6.10. Three regions are clearly distinguished. The first one for small values of ρ'_3 defined by eqn 3.32a corresponds to non-percolating fracture networks for which $K'_{\text{eff}} - 1$ is small. In contrast, the third region for large values of ρ'_3 corresponds to percolating fracture networks and $K'_{\text{eff}} - 1$ is now very large. The second region is the intermediate one where, because of the finite size effect (see Section 2.5), percolating and non-percolating networks coexist for the same value of the density. Numerical results can be summarized by distinguishing the dilute from the dense regimes.

Dilute regime

When ρ'_3 is very small, one can expect $K'_{\text{eff}} - 1$ to be proportional to ρ'_3. Of course, for larger concentrations, one should include higher order terms. Therefore, numerical data for polydisperse networks, with various exponents a, moderate densities ρ'_3 and $\sigma' = 1, 10$ and 100 suggest that

$$K'_{\text{eff}} = 1 + \kappa_d \, \rho'_3 + \kappa_d^2 \, \rho'^2_3 \qquad (6.20a)$$

6.5 *Extensions* 123

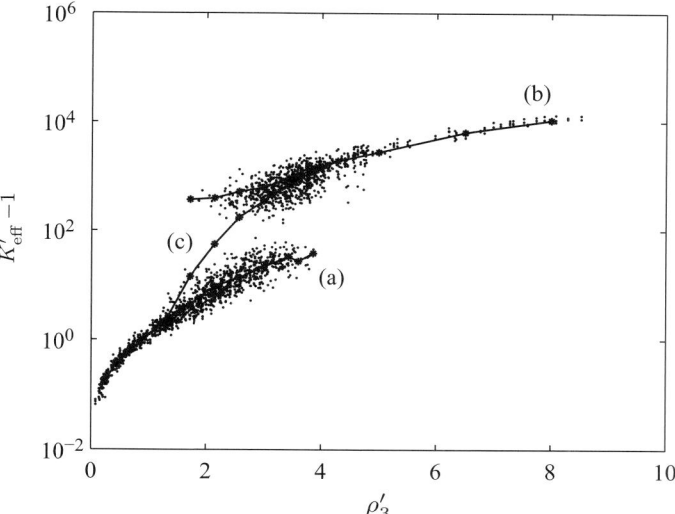

Fig. 6.10 The dimensionless permeability increment $K'_{\text{eff}} - 1$ for permeable media containing hexagonal fractures with $a = 2$, $\beta_\sigma = 0$, $\sigma' = 10^4$, $R'_m = 1/10$, $L' = 4$ and $\delta'_M = 1/4$. Dots are individual data per realization and per flow direction. Three lines join the averages of the data; for the lower densities, the first line (a) joins the averages restricted to non-percolating networks; for the larger densities, the second line (b) joins the averages restricted to percolating networks; for all the densities, a third line (c) joins the averages for all the networks.

where κ_d is deduced from

$$\kappa_d = \frac{\sigma'}{\sigma' + 3/2} \kappa_{d\infty} \qquad (6.20\text{b})$$

with $\kappa_{d\infty} = 0.335$. This value determined for hexagons is close to the theoretical result $32/9\pi^2$ for disks with $\sigma' = \infty$ and it provides a good fit for the numerical data for a wide range of fracture shapes. The form

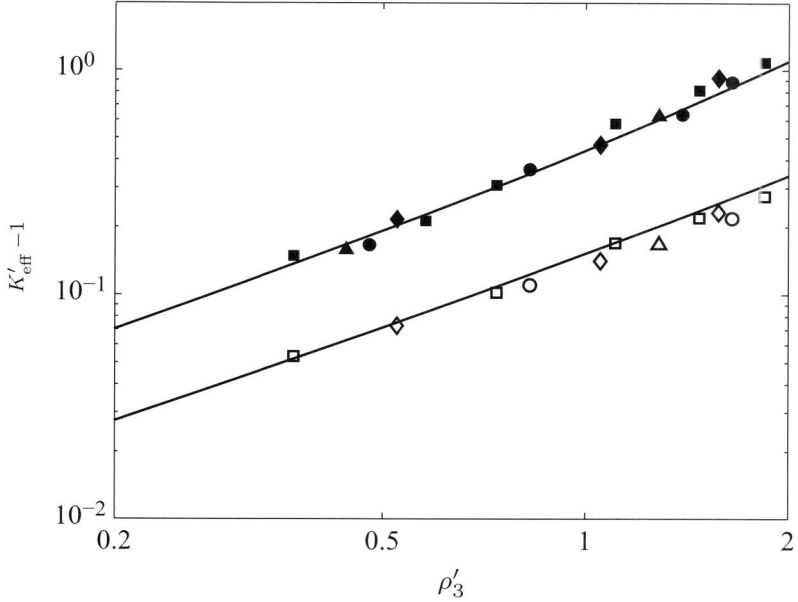

Fig. 6.11 The mean dimensionless permeability increment $K'_{\text{eff}} - 1$ for permeable media containing IOUD hexagonal fractures as a function of ρ'_3. Data are for $a = 1.5$ ($\triangle$), 2 ($\circ$), 2.5 ($\diamond$) and 2.9 ($\square$), and for $\sigma' = 1$ (open symbols) and 100 (solid symbols). The lines are the model (6.20a).

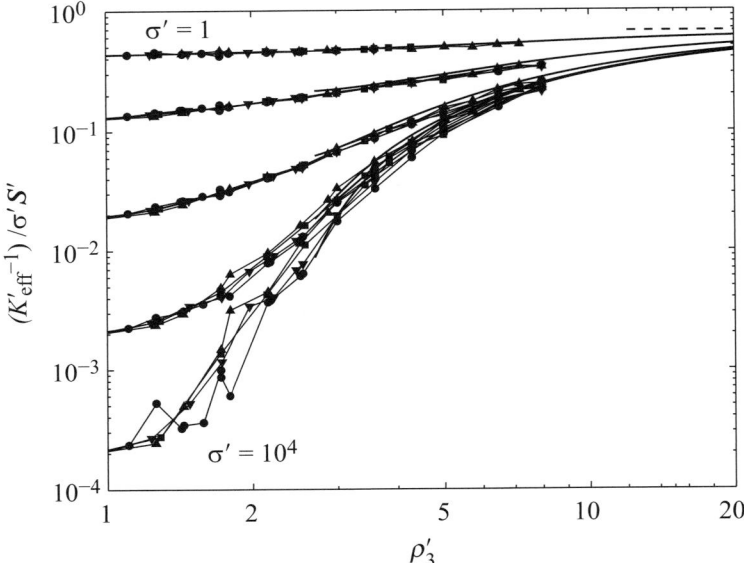

Fig. 6.12 The normalized mean permeability increments $(K'_{\text{eff}} - 1)/\sigma' \mathcal{S}'$ for permeable media containing IOUD hexagonal fractures as functions of ρ'_3. The solid lines correspond to the model eqn 6.21c. The dashed line on the right corresponds to eqn 6.12.

of eqn 6.20b is heuristic and the occurrence of the same coefficient in the first and second order terms in eqn 6.20a is fortuitous.

Equation (6.20a) is seen in Fig. 6.11 to be fairly successful in representing the permeability of moderately fractured media for $\rho'_3 < 2$, i.e. when the fracture networks never percolate. The data for $\sigma' > 100$ are not displayed because they are nearly identical to those for $\sigma' = 100$.

Dense networks

The macroscopic permeability K'_{eff} of dense networks can be obtained by the following chain of relations

$$K'_{\text{eff}} = 1 + \sigma' \, \mathcal{S}' \, K''_2 \left(\rho'_3, \sigma' \right) \tag{6.21a}$$

where $\mathcal{S}'$ is a dimensionless volumetric area weighted by transmissivity

$$\mathcal{S}' = \rho R_M \frac{\langle \sigma A \rangle}{\sigma_0} \tag{6.21b}$$

$K''_2 \left(\rho'_3, \sigma' \right)$ is expressed as

$$K''_2 \left(\rho'_3, \sigma' \right) \approx \frac{2}{3} - \frac{\frac{2}{3} - K'_2 \left(\rho'_3 \right)}{1 + \frac{7}{3} \sigma'^{-0.7}} \qquad (\, \rho'_3 \geq 4 \,) \tag{6.21c}$$

where K'_2 is defined by eqn 5.29b. This formula is valid for all the previously investigated cases, as soon as the density is large enough.

Dimensional permeability reads (with $\Delta \rho' = \rho'_3 - \rho'_{3c}$)

$$K_{\text{eff}} = K_m + \frac{2}{3} \rho \langle A\sigma \rangle \left[1 - \frac{1}{1 + \frac{7}{3} \sigma'^{-0.7}} \left(1 - \frac{\beta_K \Delta \rho'^2}{\rho' \left(1 + \beta_K \Delta \rho' \right)} \right) \right] (\rho' \geq 4) \tag{6.22}$$

where β_K (not to be confused with β_σ) is given by eqn 5.16b or by the fitted values provided in Table 5.1. The model (6.21) is compared to the numerical data in Fig. 6.12 and an excellent agreement is observed for $\rho'_3 \geq 4$. The steep rise in the curves for $\rho'_3 \approx 3$ when σ' is large corresponds to transition to percolation illustrated in Fig. 6.10. This transition has much less dramatic effect when σ' is moderate. At large densities, K''_2 tends towards $2/3$ whatever the σ' which corresponds to the result (6.12) for infinite fractures.

Applications of eqn 6.22 are proposed in Exercise 6.2.

6.5.2 Slightly compressible flows and application to pressure drawdown well tests

This subsection briefly addresses single phase, slightly compressible flow through fractured porous media and it summarizes the two papers by Bogdanov et al. (2003c) and Mourzenko et al. (2011c). Due to the specific transport properties of fractures, the flow through a naturally fractured porous medium differs drastically from that in a conventional porous medium. The key feature is that the porous matrix provides the main storage for the fluids while transport takes place mainly through the fracture system. Such situations are often described by double porosity models where the fractured porous medium is viewed as two continua representing the matrix and the fracture network (Barenblatt and Zheltov, 1960; Barenblatt et al., 1960), with exchanges governed by a coefficient known as the shape factor (Hassanzadeh and Pooladi-Darwish, 2006). In the present discrete fracture model, these exchanges are described explicitly and singular features such as the intersections of the well with the networks are shown to play a crucial role.

Consider a porous matrix with a bulk permeability K_m [L^2] which may vary in space. Darcy's law for the local seepage velocity $\overline{\boldsymbol{v}}$ and the mass conservation for slightly compressible flow can be expressed in the matrix as

$$\overline{\boldsymbol{v}} = -\frac{K_m}{\mu} \nabla p, \quad \epsilon_m C_m \frac{\partial p}{\partial t} + \nabla \cdot \overline{\boldsymbol{v}} = \delta_w J_w \quad (6.23)$$

where μ is viscosity, p is pressure, ϵ_m and C_m are the matrix porosity and total compressibility. J_w [$L^2 T^{-1}$] represents the exchanges with the well, at a location provided by the Dirac function δ_w [L^{-2}]. On this description scale, the well appears as a line with no thickness. If $\epsilon_m C_m$ is constant, a matrix pressure diffusivity D_m can be defined and eqn 6.23 can be written as a diffusion equation for pressure

$$\frac{\partial p}{\partial t} - \nabla \cdot (D_m \nabla p) = \frac{\delta_w J_w}{\epsilon_m C_m}, \quad D_m = \frac{K_m}{\mu \epsilon_m C_m} \quad (6.24)$$

The hydraulic properties of a fracture can be characterized by the two effective coefficients σ [L^3] and ω [L^{-1}], which relate the in-plane flow rate $\boldsymbol{J}$ to the surface pressure gradient $\nabla_s p$, and the seepage velocity $\overline{\boldsymbol{v}}_\perp$ of the net flow crossing the fracture to the pressure drop Δp across it by eqns 6.3 and 6.4.

To illustrate this, the fracture can be viewed as a plane channel of aperture b, filled with a porous material with permeability K_f, porosity ϵ_f and total compressibility C_f. Then, σ and ω are given by eqn 6.6a. The continuity equation for the flow through a fracture reads as

$$b\,\epsilon_f\,C_f\frac{\partial p}{\partial t} + \nabla_s \cdot \boldsymbol{J} = (\overline{\boldsymbol{v}}^- - \overline{\boldsymbol{v}}^+)\cdot \boldsymbol{n} + b\delta_w J_w \qquad (6.25)$$

where $\overline{\boldsymbol{v}}^\pm$ is the seepage velocity in the matrix on either side of the fracture, given by (6.23). Again, if $b\epsilon_f C_f$ is constant in a fracture, a pressure diffusivity D_f can be defined, and eqn (6.25) can be written as

$$\frac{\partial p}{\partial t} - \nabla_s \cdot (D_f \nabla_s p) = \frac{(\overline{\boldsymbol{v}}^- - \overline{\boldsymbol{v}}^+)\cdot \boldsymbol{n}}{b\,\epsilon_f\,C_f} + \frac{\delta_w J_w}{\epsilon_f\,C_f}, \qquad D_f = \frac{\sigma}{b\,\mu\epsilon_f C_f} \qquad (6.26)$$

The assumptions of uniform $\epsilon_m C_m$ and $b\epsilon_f C_f$ used to write down the diffusion equations (6.24, 6.26) are by no means a requirement for the numerical model, since the discretized equations are really based on general equations (6.23, 6.25).

The model can accommodate the usual boundary conditions in reservoir engineering, namely of a Dirichlet type, with an imposed far-field pressure p_∞, or of a Neumann type, with no-flux conditions at the boundaries of a closed reservoir. For systematic studies of the interactions of a single well with a fractured formation, a simple geometrical setting is used, with a parallelepipedic cell. The well is set along a vertical line throughout the cell, and spatial periodicity is applied along the x- and y-directions. Hence, this situation corresponds to a periodic square array of infinite vertical wells, or to a single well in a closed reservoir. All the simulations reported here correspond to a pressure drawdown test, i.e. to a well producing at constant rate Q_w from a field initially at rest. The initial pressure is arbitrary, and can be taken as $p_0 = 0$ with no loss of generality.

The fracture network and the space located in between the fractures are successively triangulated by the advancing front technique described in Section 6.3.1. The well is treated as a 1D highly conductive system which is superimposed on the medium. Time derivatives are discretized to first order, in a fully implicit formulation.

Application of the previous developments was made to a fractured porous medium with a dimensionless density ρ' equal to 6. The cell size is $L = 6R$; the fractures are plane hexagons with two dimensionless transmissivities $\sigma' = 10^2$ and 10^3.

Typical results are shown in Fig. 6.13 for a well radius $r_w = 0.01R$ in terms of the dimensionless quantities

$$p'_w = \frac{4\pi K_m}{\mu J_w} p_w, \quad t' = \frac{D_m t}{R^2}, \quad \Pi'_w = \frac{\mathrm{d}p'_w}{\mathrm{d}\ln t'} \qquad (6.27)$$

where p_w is the dimensional well pressure. In this case, the actual number of intersections of the well with the fracture network is equal to the expected value $N_i = 6$ (see eqn 3.14a). The density of intersections of

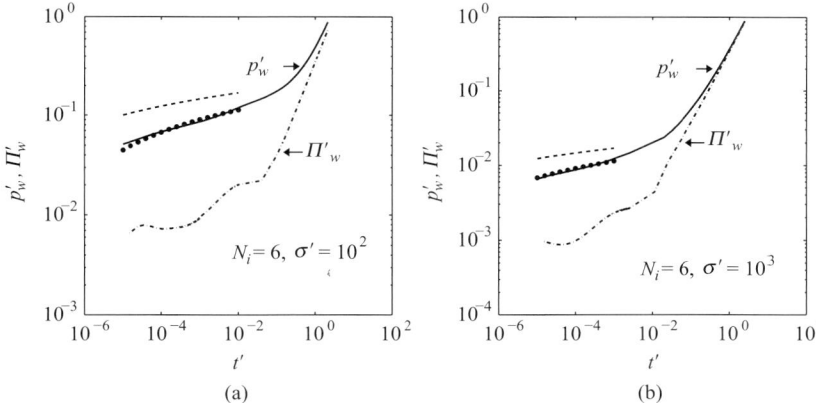

Fig. 6.13 Pressure p'_w (———) and its derivative Π'_w ($-\cdot-\cdot-$) for a pressure drawdown test, for a fracture network with $\rho' = 6$ in a unit cell with size $L = 6R$. The well intersects six fractures. Data are for $r_w = 0.01R$, $b' = 10^{-3}$; $\sigma' = 10^2$ (a) and $\sigma' = 10^3$ (b). The analytical solution by Boulton and Streltsova (1977) for a spacing equal to the mean value $S_i = R$ is given for the theoretical ($- - - -$) and corrected ($\cdots\cdots$) D_f (see Bogdanov et al., 2003c).

the well with the network is obviously an important parameter. The four usual flow regimes can be observed, i.e. transient and pseudo-steady fracture flows, followed by transient and pseudo-steady global flows. The decrease in the derivative Π'_w at $t' \sim 10^{-4}$ is commonly observed, and corresponds to the spread of the pressure wave in the ramified network, with decreasing hydraulic resistance.

Systematic simulations of drawdown tests have been run throughout a wide range of network density ρ' and fracture transmissivity σ' by Bogdanov et al. (2003c). In the late pseudo-steady stage, the well pressure always increases linearly with time. The results can be analyzed by comparison with the pressure response of a well in a homogeneous medium with permeability K_g and storage coefficient $\epsilon_g C_g$ (cf. Earlougher, 1977)

$$p_w(t) = \frac{J_w}{\epsilon_g C_g \mathcal{A}} t + \frac{\mu J_w}{4\pi K_g}\left(\ln\frac{\mathcal{A}}{r_w^2} + \ln\frac{2.2458}{C_A}\right) \quad (6.28)$$

where $\mathcal{A}$ is the drainage area; the shape factor C_A depends on the reservoir geometry and on the location of the well. Bogdanov et al. (2003c) showed that as a rule, K_g is significantly larger than the fractured medium effective permeability K_{eff} and strongly depends on the number of well/fractures intersections. Hence, K_g cannot be regarded as an intrinsic property of the fractured medium, since it depends on boundary conditions.

On physical grounds, the well pressure response can be expected to depend on the far-field effective permeability K_{eff} of the medium, and on the particular interactions of the well with the fracture network in a given situation. This can be modeled in a first approximation by introducing a skin factor S_k in eqn 6.28

$$p_w(t) = \frac{J_w}{\epsilon_g C_g \mathcal{A}} t + \frac{\mu J_w}{4\pi K_{\text{eff}}}\left(\ln\frac{\mathcal{A}}{r_w^2} + \ln\frac{2.2458}{C_A} + 2S_k\right) \quad (6.29)$$

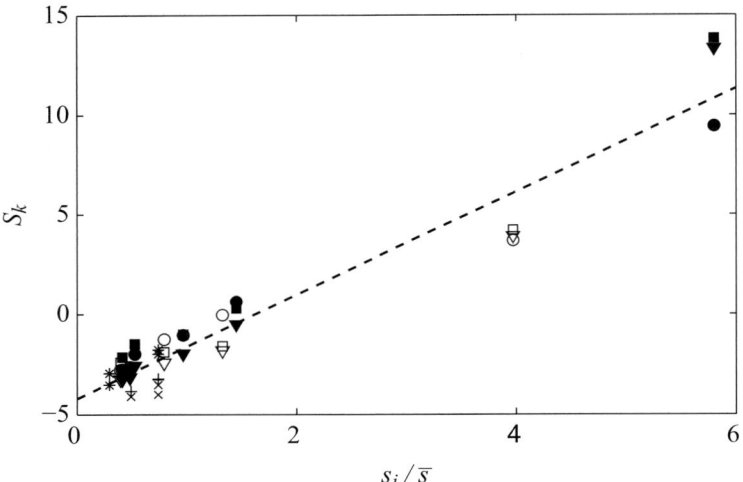

Fig. 6.14 Skin factor S_k as a function of the spacing $s_j/\bar{s}$. Data are given for a well with a radius $r_w = 0.01R$ in a randomly fractured medium, with various densities ρ' and fracture conductivities σ'. Values are: ● ($\rho' = 6$, $\sigma' = 10$), ▼ ($\rho' = 6$, $\sigma' = 100$), ■ ($\rho' = 6$, $\sigma' = 1000$), ○ ($\rho' = 4$, $\sigma' = 10$), ▽ ($\rho' = 4$, $\sigma' = 100$), □ ($\rho' = 4$, $\sigma' = 1000$), ∗ ($\rho' = 1.5$, $\sigma' = 10$), + ($\rho' = 1.5$, $\sigma' = 100$), × ($\rho' = 1.5$, $\sigma' = 1000$). The broken line correspond to the fit (6.30).

S_k is positive if permeability near the well is smaller than in the surrounding medium.

The skin factor is plotted in Fig. 6.14 as a function of the ratio $s_j/\bar{s}$ of the mean spacing s_j between the well/fracture intersections and of its expected value $\bar{s}$ (see eqn 3.14e). A linear fit is possible, which provides a guide line to predict the performances of a well, given its connectivity with the fracture network, for highly conducting and well-connected fractures

$$S_k = 2.5(\frac{s_j}{\bar{s}} - 1) - 1.6 \qquad (r = 0.963) \qquad (6.30)$$

where r is the correlation coefficient of the linear fit.

Exercises

(6.1) (i) In order to clarify the physical meaning of σ, compare the flow rates Q_m and Q_f passing through a cubic block of porous medium $R \times R \times R$ and through a square fracture $R \times R$. Both are submitted to the same overall pressure gradient $\overline{\nabla p}$.

(ii) Determine σ and ω for a plane fracture of constant aperture b and filled with a porous medium of permeability K_f. In other words, prove eqn 6.6.

(6.2) (i) Use the master curves of Fig. 6.9 and determine K'_{eff} for the three networks given in Exercise 3.7; $\sigma' = 10^{-6}$, 10^2.

(ii) Apply the simple approximation provided in Section 6.2.3 and the more precise formula (6.22).

(iii) Comment on the results.

Two-phase flow through fractured porous media

7.1 Introduction	129
7.2 Local equations on the Darcy scale	129
7.3 Numerical approach	133
7.4 Regular fracture networks	134
7.5 Isotropic and homogeneous fractured porous media	138
7.6 Comparison with a capillary dominated model	143
Exercises	145

7.1 Introduction

This chapter is a short introduction to the field of two-phase flows through fractured porous media which is extremely important for industrial applications. For instance, air is present in the capillary fringe above free aquifers. Moreover, there is water present in almost all oil fields and on top of this there is often gas, and therefore three phases can be present. Theoretically speaking, there are many problems faced since the form of the equations on the Darcy scale that we are going to use are not established on a firm theoretical background; in addition, these equations need to be supplemented by some *constitutive relations*; these theoretical difficulties are not addressed here.

The main objective of this chapter is to upscale the local properties, i.e. to calculate the relative permeabilities and the capillary pressure on the large scale of the fractured porous medium. This chapter is based on the paper by Bogdanov *et al.* (2003b) to which the reader is referred for further details and references.

This chapter is organized as follows. The classical extensions of the Darcy equation to two-phase flow are presented in Section 7.2, together with the standard constitutive equations of van Genuchten (1980); some general remarks are made on the restrictions of the present model and its possible generalizations. The numerical approach is briefly addressed in Section 7.3. Then, some simple applications to regular fracture networks are made in order to illustrate specific aspects of two-phase flows. Finally, the important case of isotropic and homogeneous fractured porous media (i.e. I^2OUD) is addressed in Section 7.5; a simple capillary dominated model is exposed in order to approximate the numerical results.

It should be noted that the nomenclature in this chapter is specific and is listed separately.

7.2 Local equations on the Darcy scale

7.2.1 Conservation equations

First recall that the governing flow equations are written at an intermediate scale, which is small compared to the fracture extension but large compared to the typical pore size in the matrix and to the typical fracture aperture. Hence, they result from the homogenization of the microscopic Stokes equations, and the standard requirements of statistical homogeneity for this preliminary upscaling are meant to be fulfilled.

The subscript $i = w, n$ refers to wetting and non-wetting fluids, respectively. The subscript $j = m, f$ refers to the porous medium and fractures, respectively. In order to keep the notations as simple as possible, the value of the subscript j is often omitted in this chapter when no ambiguity is present.

Generally speaking, the saturation S_{ij} of fluid i $(i = w, n)$ is the portion of void space occupied by fluid i in the medium j $(j = m, f)$. Therefore, four saturations can be introduced in a fractured porous medium.

The porous rock matrix will have a porosity ϵ_m and a bulk permeability K_m $[L^2]$ that can vary with space. The flow in the matrix is described by a generalized Darcy law for each phase, with relative permeabilities $K_{r,i}$ $(i = w, n)$. The local seepage velocities $\overline{v}_i$ are expressed by

$$\overline{v}_i = -\frac{K_m \, K_{r,i}}{\mu_i} \nabla (p_i - \rho_i g z) \qquad (i = w, n) \qquad (7.1a)$$

where μ_i is viscosity, ρ_i density and p_i pressure for fluid i. g denotes gravity and z the vertical axis oriented downwards. For the sake of concision, denote Φ_i the potential $p_i - \rho_i g z$. Then, eqn 7.1a reads

$$\overline{v}_i = -\frac{K_m \, K_{r,i}}{\mu_i} \nabla \Phi_i \qquad (i = w, n) \qquad (7.1b)$$

The fluids are considered as incompressible. Hence, two continuity equations and a global condition on the saturations S_{im} $(i = w, n)$ can be written for the porous medium

$$S_{nm} + S_{wm} = 1 \qquad (7.2a)$$

$$\epsilon_m \frac{\partial S_{im}}{\partial t} + \nabla \cdot \overline{v}_i = 0 \qquad (i = w, n) \qquad (7.2b)$$

Equations similar to 7.1 and 7.2 are applied to the flow through the fractures. The in-plane flow rates $\boldsymbol{J}_i$ per unit width are related to the surface pressure gradients $\nabla_s p_i$ via two-dimensional generalized Darcy laws

$$\boldsymbol{J}_i = -\frac{\sigma \, \sigma_{r,i}}{\mu_i} \nabla_s \Phi_i \qquad (i = w, n) \qquad (7.3)$$

where σ is the absolute fracture transmissivity (cf. eqn 5.1) which can be position-dependent and fracture-dependent; $\sigma_{r,i}$ $(i = w, n)$ are the relative fracture transmissivities.

In this chapter, the fractures are assumed to create a negligible hydrodynamic resistance to flow normal to their plane, i.e. ω defined by eqn 6.4 is equal to 0. Hence, the pressures p_i $(i = w, n)$ and the potentials Φ_i are continuous across the fractures.

Conservation equations which are similar to eqn 7.2 could be written for the fractures, which should include exchange terms with the surrounding matrix. In view of the finite volume scheme used for the numerical solution, it is more convenient to write a global conservation equation, that accounts for both matrix and fracture flow in a control volume. Suppose that volume Ω, with boundary $\partial\Omega$, contains part of one or several fractures, denoted by F. By applying the divergence theorem, conservation of phase i in Ω can be written as

$$\int_{\Omega-F} \epsilon_m \frac{\partial S_{im}}{\partial t} dv + \int_{\Omega \cap F} \epsilon_f \frac{\partial S_{if}}{\partial t} dv + \int_{\partial\Omega - F} \boldsymbol{n}.\overline{\boldsymbol{v}}_i ds$$
$$+ \int_{\partial\Omega \cap F} \boldsymbol{n}.\boldsymbol{J}_i dl = 0 \qquad (i = w, n) \qquad (7.4)$$

where $\boldsymbol{n}$ is the unit vector normal to $\partial\Omega$ and ϵ_f is the porosity of the fracture filling material. The volume of the fractures in F is supposed to be negligible compared to the pore volume in the matrix in $\Omega - F$. Therefore, eqn 7.4 can be simplified as

$$\int_{\Omega} \epsilon_m \frac{\partial S_{im}}{\partial t} dv + \int_{\partial\Omega} \boldsymbol{n}.\overline{\boldsymbol{v}}_i ds + \int_{\partial\Omega \cap F} \boldsymbol{n}.\boldsymbol{J}_i dl = 0 \qquad (i = w, n) \qquad (7.5)$$

Thus, the fractures introduce a singular contribution to the mass balance equation (7.2b).

The flow equations can be rewritten in terms of one of the potentials and of the capillary pressure, thanks to (7.1)–(7.3).

7.2.2 Constitutive equations

Constitutive equations are required for the closure of the set of transport equations in the previous subsection. More precisely, these equations link capillary pressures, relative permeabilities and transmissivities to the fluid saturations in the fractures and in the matrix.

First, the stress balance at the fluid interface at a microscopic scale has to be taken into account. Due to interfacial tension, a pressure jump p_c takes place across the interface, which is called the *capillary pressure*

$$p_c = p_n - p_w = \Phi_n - \Phi_w + \Delta\rho g z, \quad \Delta\rho = \rho_n - \rho_w \qquad (7.6)$$

p_c is usually related to S_w and the most widely used relation was proposed by van Genuchten (1980)

$$S_w = \left[1 + \left(\frac{p_c}{p_0}\right)^n\right]^{\frac{1-n}{n}} \qquad (7.7)$$

where p_0 is a characteristic pressure; n is an index. Typical values of n range from 1 to 4. Estimates of p_0 are given below. These equations can be written for the porous medium and for the fractures in which cases the subscript $j = (m, f)$ should be added.

Recall that owing to the local equilibrium hypothesis, pressures p_n and p_w and the capillary pressure p_c are continuous; they are equal in a fracture and in the matrix rock along its surface. However, parameters p_0 and n are generally different in both domains. In particular, at the microscopic scale, the pressure jump across the interface between the two fluids is proportional to the surface tension γ and inversely proportional to the meniscus radius, which is of the order of a typical pore size. Therefore, one can expect

$$p_0 \propto \frac{\gamma}{\sqrt{K}}, \qquad \frac{p_{0,f}}{p_{0,m}} \approx \sqrt{\frac{K_m}{K_f}} = \kappa \qquad (7.8)$$

Equation (7.7) is applied to both the rock matrix and the fractures. Unless otherwise stated, the computations in the following correspond to $p_{0,f} = \kappa p_{0,m}$.

On the other hand, the relative permeabilities and transmissivities appearing in the generalized Darcy equations (7.1 and 7.3) also depend on fluid saturations. Many models have been considered for the porous media, but again the most widely used model for the relative permeability of the wetting phase was proposed by Mualem (1976) and later by van Genuchten (1980)

$$K_{r,w} = S_{wm}^{1/2} \left[1 - \left(1 - S_{wm}^{\frac{n}{n-1}} \right)^{\frac{n-1}{n}} \right]^2 \qquad (7.9)$$

Note that eqns 7.7 and 7.9 suppose that S_{wm} (S_{wf}, resp.) can vary over the whole range from 0 to 1. If its practical variations are limited by irreducible and maximal values S_{wmr} and S_{wms} (S_{wfr} and S_{wfs}, resp.), eqns 7.7 and 7.9 are generally written in terms of the effective saturation $\tilde{S}_{wm}$ ($\tilde{S}_{wf}$, resp.)

$$\tilde{S}_{wm} = \frac{S_{wm} - S_{wmr}}{S_{wms} - S_{wmr}}, \quad \tilde{S}_{wf} = \frac{S_{wf} - S_{wfr}}{S_{wfs} - S_{wfr}} \qquad (7.10)$$

Residual saturations are not considered in the present simulations, but this could easily be done. Again the subscript $j = (m, f)$ should be added to several of the previous relations.

The relative permeability for the non-wetting phase is also sometimes modeled according to eqn 7.9, with $K_{r,w}$ and S_w replaced by $K_{r,n}$ and S_n, respectively. However, the relative permeability curves for the two phases are generally not mirror images of one another. Thus, a different model was used here, which is discussed below.

Two-phase flows in fractures have given rise to comparatively fewer experimental studies than three-dimensional porous media, but a few references can be found in the literature which are reviewed by Bogdanov et al. (2003b). In the present simulations, $\sigma_{r,n}$ is expressed as

$$\sigma_{r,n} = S_{nf}^q \qquad (7.11)$$

with the exponent q equal to 2. $K_{r,n}$ in the rock matrix is provided by the same equation as (7.11), but in terms of S_{nm}. Moreover, $\sigma_{r,w}$ is described by an equation similar to (7.9). It can be noted that in most of the situations considered here, the wetting phase saturation in the fractures is very small since their aperture is usually much larger than the pore size; therefore, $\sigma_{r,w}$ is very small and $\sigma_{r,n}$ is of the order of unity for any reasonable choice of constitutive equations.

In summary, the capillary pressure p_c, the relative permeabilities and transmissivities for the wetting fluid $K_{r,w}$ and $\sigma_{r,w}$ and the relative permeabilities and transmissivities for the non-wetting fluid $K_{r,n}$ and $\sigma_{r,n}$ are described both in the matrix and in the fractures by eqns 7.7, 7.9 and 7.11, respectively. Applications of the van Genuchten relations are proposed in Exercise 7.1.

At equilibrium, because of the capillary pressure, the non-wetting fluid is preferably in the fractures; this simple remark has major consequences for the relative permeabilities as detailed in Section 7.5. In the examples which are going to be presented, capillary forces are dominant over viscous forces.

Of course, the equations are not used as such, and as usual in this field, they are made dimensionless. Bogdanov *et al.* (2003b) provide a complete presentation which can be summarized as follows. The major units are a characteristic length scale L (which is defined specifically in each case), a pressure unit $p_{0,m}$ and a permeability unit K_m. Some of the dimensionless numbers are

$$\sigma' = \frac{\sigma}{LK_m}; \quad \kappa = \frac{p_{0,f}}{p_{0,m}}; \quad \text{exponents} \quad n, q \qquad (7.12)$$

Another important dimensionless parameter is the capillary number C which compares the magnitudes of the viscous and interfacial stresses. C is defined as

$$C = \frac{\mu U}{\gamma} \qquad (7.13)$$

where U is a typical fluid velocity. This number is generally very small for underground flows (typically less than 10^{-6}).

7.3 Numerical approach

7.3.1 Spatial discretization

The rock matrix is represented by tetrahedral volume elements, and the fractures by triangular surface elements. The transport coefficients K_m and σ, as well as the porosity ϵ_m, are considered as uniform throughout these elements.

The non-wetting phase potential Φ_n and the capillary pressure p_c are evaluated at the mesh points located on the vertices of the tetrahedra and triangles. Since the fractures are viewed as vanishingly thin, empty or very permeable layers, there is no pressure jump between two points facing each other on the two opposite sides of a fracture. Thus, single values of Φ_n and p_c can be used in the numerical formulation per vertex of fracture element.

A finite volume formulation of the problem is obtained by applying balance equations such as eqn 7.5 over control volumes Ω surrounding each of the mesh points as shown in Fig. 6.5a.

Fig. 7.1 Homogeneous rock matrix containing an array of infinite parallel fractures. (a) Notations. (b) Relative permeabilities as functions of the saturation of the wetting phase. The solid lines are the curves for the matrix and for the fractures, which are identical, with $n_m = n_f = q = 2$. The other lines correspond to eqn 7.17 for $\sigma' = 1/4$, $\kappa = 10^{-3/2}$ ($-\cdot-\cdot-$) and $\sigma' = 25$, $\kappa = 10^{-5/2}$ ($----$). The symbols are results of numerical simulations, for $\overline{K}_{r,n}$ (□) and $\overline{K}_{r,w}$ (○).

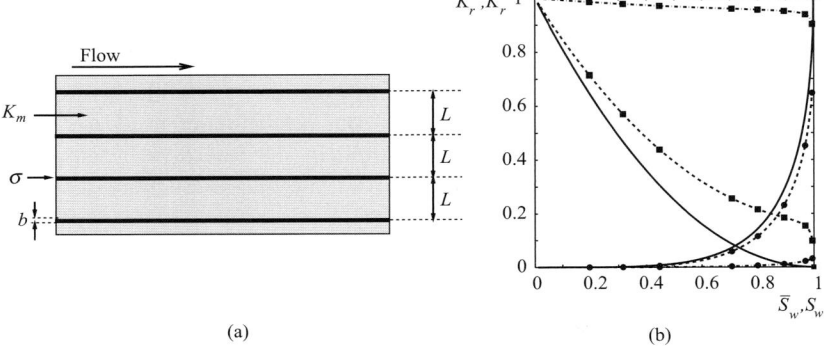

An analogous numerical approach was adopted by Matthai *et al.* (2005).

7.3.2 Resolution of the equations

The strong non-linearities of the coefficients in the equations presented in Section 7.2 require an implicit time formulation. The one used here is an extension of the modified Picard scheme described by Celia *et al.* (1990) for the solution to Richards' equation. The main idea is linearization of the time derivative in a mass conservative form. Further details of this specific application are found in Bogdanov *et al.* (2003b).

7.4 Regular fracture networks

7.4.1 Multiple families of parallel plane fractures

Consider the situation sketched in Fig. 7.1a of a porous matrix that contains an array of infinite parallel plane fractures, with a spacing L which is considered as the unit length. A flow is imposed by a pressure gradient which is parallel to the fractures. Matrix permeability is K_m, fracture transmissivity is σ and fracture aperture verifies $b \ll L$. The macroscopic absolute permeability of this fractured medium, in the direction of the fractures, is

$$K_{\text{eff}} = \frac{(L-b)K_m + \sigma}{L} \approx K_m(1+\sigma') \qquad (7.14)$$

where σ' is provided by eqn 7.12.

Capillary pressure is supposed to be uniform initially and equal to p_c, and it obviously remains so for a stationary flow parallel to the fractures. The wetting fluid saturations in the matrix and in the fractures, S_{wm} and S_{wf} differ because of different capillary properties; S_{wm} and S_{wf} are related to each other by the equality of the capillary pressures. The mean global saturation $\overline{S}_w$ is dominated by the matrix saturation

$$\overline{S}_w = \frac{(L-b)\epsilon_m S_{wm} + b\epsilon_f S_{wf}}{(L-b)\epsilon_m + b\epsilon_f} \approx S_{wm} \qquad (7.15)$$

phase distributions with bands which are normal or parallel to the applied pressure gradient $\overline{\overline{\nabla p}}$.

In all cases, an identical steady regime is reached, where saturation is not uniform. A precise estimation of the time necessary to reach equilibrium is provided by Bogdanov *et al.* (2003b). S_{wm} ranges from about 0.32 to 0.43, and it is different at the inlet and outlet sides of the fractures. The disturbances in the saturation field introduced by the presence of the fractures during a steady flow with respect to the rest state are due to the different capillary properties of the fractures and rock matrix (see eqn 7.8). They are observed in all our simulations, and increase with the fracture transmissivity σ' and with the mean flow rate (or pressure gradient).

This example will conclude with two important remarks. The first one is that, as expected, the non-wetting fluid is preferably in the fractures and this fact has a significant impact on the relative permeability of this phase.

The second remark is even more important. It is crucial to obtain the same result, whatever the initial conditions, in order to define macroscopic properties such as relative permeabilities. This independence of the final results on the initial conditions is called *ergodicity*. This term has already been used for spatial fields in Subsection 2.6.2 with a different meaning.

Therefore, the equations are ergodic when there are no residual saturations. It is easy to realize that as soon as there is a non-zero residual saturation of either phase, one can devise initial conditions of saturation configurations in such a way as to obtain very different results.

Another independent cause of non-ergodicity is the existence of hysteresis in the constitutive laws.

7.5.2 Steady state macroscopic properties

The previous example has shown that identical saturation fields and phase flow rates are eventually reached when a fractured medium is submitted to a macroscopic pressure gradient $\overline{\overline{\nabla p}}$, starting from very different initial phase distributions. Therefore, it is possible to define steady state macroscopic phase relative permeabilities for this medium at a given mean saturation $\overline{\overline{S}}_w$. These relative permeabilities $\overline{K}_{r,i}$ are intrinsic in that they do not depend on the initial conditions; Bogdanov *et al.* (2003b) showed that they depend neither on the magnitude of the applied macroscopic pressure gradient, nor on the viscosity ratio, at least in a reasonable range.

Thus, for the steady flow of two given fluids in a fractured medium, it is possible to relate the mean global phase flow rates $\overline{\overline{v}}_i$ to the phase pressure gradients using a generalized Darcy law of the form (7.1)

$$\overline{\overline{v}}_i = -\frac{K_{\text{eff}}\,\overline{K}_{r,i}}{\mu_i}\left(\overline{\overline{\nabla p_i}} - \rho_i g \mathbf{e}_z\right) \qquad (i = w, n) \qquad (7.20)$$

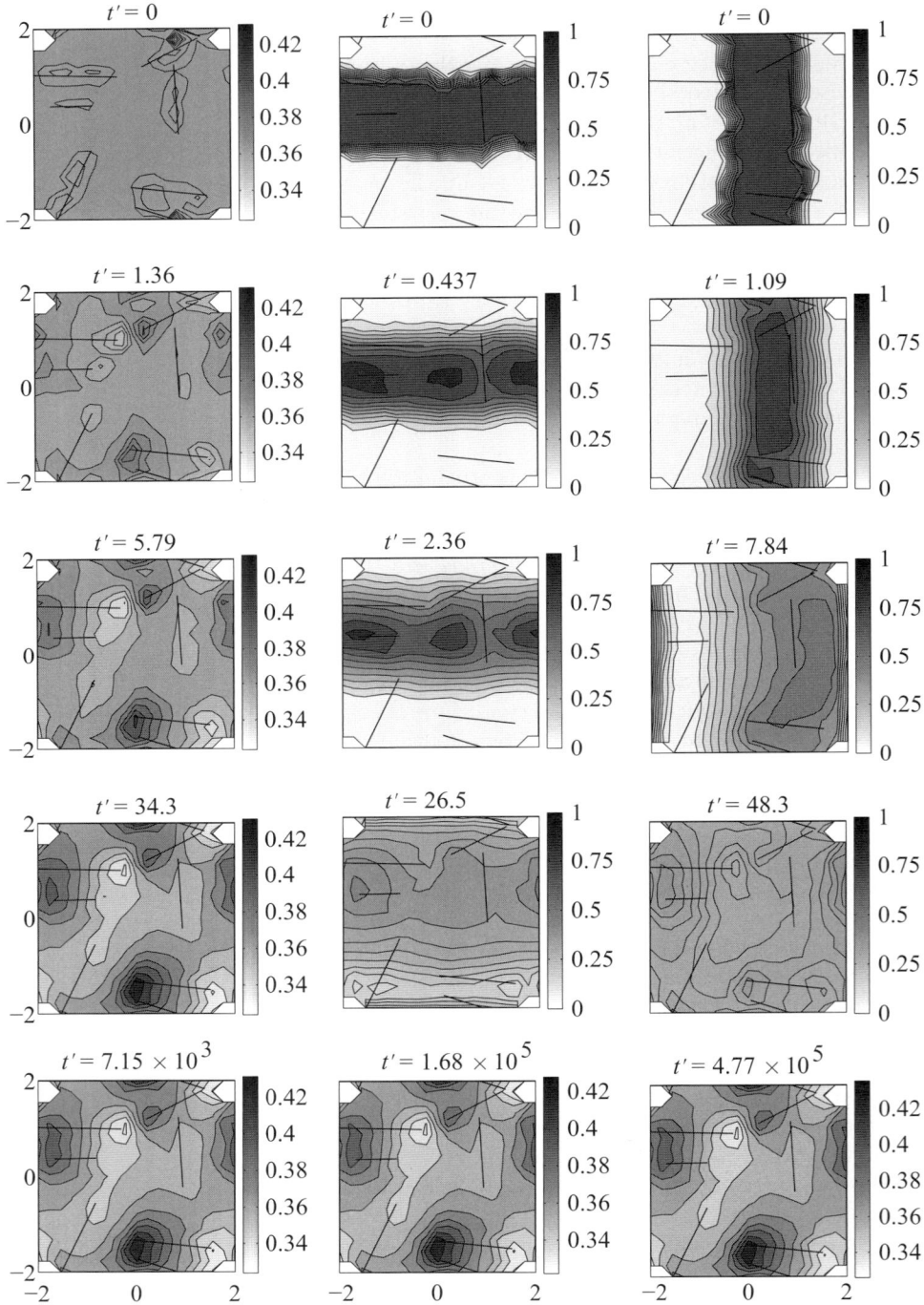

Fig. 7.5 Wetting phase saturation at various times (top to bottom), in the plane Π marked in Fig. 7.4a. The mean flow is oriented from the left to the right of the figure. The mean saturation is always $\overline{S}_w = 0.371$. Each column corresponds to the evolution of the saturation which starts from the initial condition shown at the top of the column.

where K_{eff} is the macroscopic absolute permeability of the fractured porous medium as discussed in Chapter 6, and e_z is the unit vector of the vertical z-axis which is directed downwards. We should emphasise that only I²OUD fracture networks are considered in this chapter.

Equation 7.20 is the first step towards the upscaling of the steady state two-phase flow problem. It corresponds to the transport equation (7.1) in a homogeneous material. The conservation equation (7.2) also applies to $\overline{S}_i$ and $\overline{\overline{v}}_i$. Finally, the data for the relative permeabilities $\overline{K}_{r,i}$ from the present numerical calculations are the macroscale counterpart of the constitutive equation (7.9). The final missing element is an upscaled capillary pressure-saturation relationship. This topic will be discussed at the end of this subsection.

In view of the large number of parameters, which include the rock and fractures geometry and transport properties, the fluid characteristics and coefficients in the constitutive equations, in addition to the initial and boundary conditions, we chose to study only the influence of the mean saturation $\overline{S}_w$ on $\overline{K}_{r,i}$, for two fracture densities and a single typical value of the other parameters. Calculations are made for $r_m = n_f = q = 2$, $\mu_n/\mu_w = 10$, $\rho_n = \rho_w$; fracture transmissivity is $\sigma' = 1$ with $\kappa = 10^{-3/2}$.

However, we consider samples of size $L = 4R$ containing either 16 or 32 randomly located fractures. An example with 32 fractures is shown in Fig. 7.4c. $\mathcal{N}_r = 9$ realizations are generated in each case, and the flow equations are solved with a pressure gradient of magnitude $p_{0,m}/R$ set along the x-, y- and z-axes, successively. In the first case, the percolation probability of the fracture network in a prescribed direction is numerically around 20%, whereas in the latter, it is around 80%. The examples displayed in Figs 7.4b and 7.4c do not percolate and percolate, respectively.

Computations are run starting from initial rest state equilibrium conditions, i.e. with a uniform capillary pressure corresponding to various mean saturations $\overline{S}_w$ in the range 0.1∼0.9, until convergence of the saturation field. In addition, a single-phase calculation was performed in order to determine the sample's absolute permeability. The relative permeabilities $\overline{K}_{r,i}$ for each case are deduced from the phase flow rates via eqn 7.20.

Results are shown in Figs 7.6a and 7.6b, for the 16- and 32-fracture samples, respectively. The symbols correspond to the statistical averages over 27 calculations, and the error bars to the full range of variation in the individual data. The solid lines are the relative transmissivities and permeabilities for the fractures and for the rock matrix, which are identical functions of saturation, in the present case.

In spite of the difference in percolation probability between the two cases, the general aspects of the results are similar. The presence of fractures increases the relative permeability for the non-wetting phase and decreases the relative permeability for the wetting phase, compared with the intact matrix material. However, the amplitude of these variations is greater for the denser fracture networks.

The strongest effects are observed for the largest saturations, and for the non-wetting fluid permeability $\overline{K}_{r,n}$. This is a consequence of the different capillary functions of the fractures and rock matrix (see eqn 7.8). For the same value of p_c, the non-wetting phase saturation is much larger in the fractures than in the surrounding matrix, and the relative transmissivity $\sigma_{r,n}$ is larger than $K_{r,n}$. Thus, the fractures are preferential paths for the non-wetting phase.

Conversely, $\overline{K}_{r,w}$ is smaller than $K_{r,w}$ in the rock matrix, but this is mostly a consequence of the increase in the absolute permeability induced by the presence of the fractures, $K_{\text{eff}} > K_m$. More precisely, for a given saturation, with or without fractures,

$$K_{\text{eff}}\, \overline{K}_{r,w} \approx \text{const} \tag{7.21}$$

Let us finally consider the macroscopic capillary pressure–saturation relationship. The mean saturation $\overline{S}_w$ corresponds in a rest state to a capillary pressure $p_{c,r}$. Since the interstitial volume in the medium is widely dominated by the pore volume in the rock matrix, $p_{c,r}$ is related to $\overline{S}_w$ by the law (7.7) for the matrix, with $n = n_m$ and $p_0 = p_{0,m}$. When a steady flow takes place through the fractured medium, the saturation and capillary pressure fields are not uniform (see Fig. 7.5). The volume average of the fluctuations of S_w is zero, since $\overline{S}_w$ is conserved. However, due to the nonlinearity of eqn 7.7, the volume average $\overline{p}_c$ is not necessarily equal to $p_{c,r}$. It was calculated for the 27 steady states obtained in the two types of fractured media. It appears that the difference between $\overline{p}_c$ and $p_{c,r}$ is negligible. Hence, the volume averaged capillary pressure does not differ between rest state and steady flow.

Recall that in the present case the global volume average is equivalent to an average over the matrix only, since the fracture volume is negligible. This equivalence may not be valid if the matrix contains three-dimensional heterogeneities, such as regions of a more permeable material. In this case, the concentration of the non-wetting fluid in the

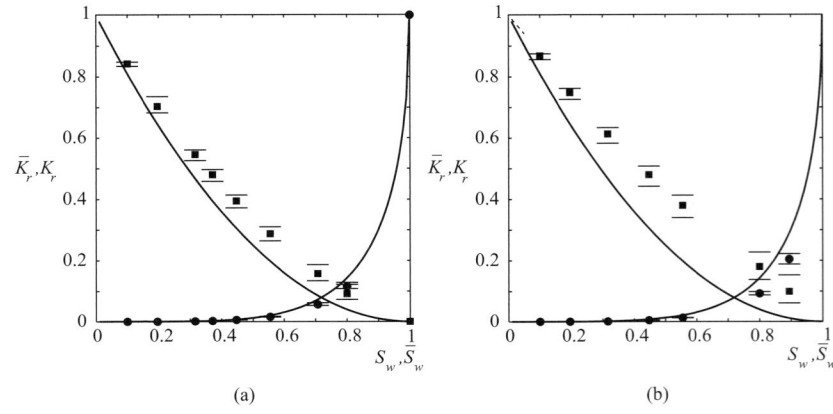

Fig. 7.6 Macroscopic relative permeabilities $\overline{K}_{r,i}$ as functions of the mean saturation $\overline{S}_w$. Data are for samples containing 16 (a) or 32 (b) hexagonal fractures. The cell size is $4R$. The fractures have a transmissivity $\sigma' = 1$, and $\kappa = 10^{-3/2}$. The fluids have equal densities. The symbols are the averages of $\overline{K}_{r,n}$ (□) and $\overline{K}_{r,w}$ (○) over 27 calculations, conducted in $\mathcal{N}_r = 9$ random realizations; the pressure gradient is equal to $p_{0,m}/R$ and is set along the x-, y- and z-axes. The horizontal lines show the full variation range of the individual data. The solid lines are the relative transmissivities and permeabilities for the fractures and for the rock matrix, with $n_m = n_f = q = 2$.

permeable region would induce a noticeable decrease of its saturation in the matrix.

7.5.3 Influence of the parameters

A systematic study of the parameters was conducted by Bogdanov et al. (2003b) for the two fracture networks analyzed in the previous subsection and displayed in Fig. 7.4. Only the major conclusions are given here. It should be remembered that no study has been conducted so far for two-phase flows in fractured porous media by varying the fracture density in a systematic way.

From the two networks which have been analyzed, it can be said that when σ' increases, the wetting fluid relative permeability $\overline{K}_{r,w}$ decreases, but this is merely a consequence of the increase of K_{eff}; indeed, the product $K_{\text{eff}} \overline{K}_{r,w}$ remains nearly constant. Simultaneously, there is an increase of the non-wetting fluid saturation in the fractures.

Denote by G the dimensionless macroscopic pressure gradient, i.e.

$$|\overline{\overline{\nabla p}}| = G \, \frac{p_{0,m}}{R} \qquad (7.22)$$

It was decreased to 0.1 or increased to 10 in a few cases. Its influence on the relative permeabilities is less than 20% in all the cases. A few calculations run with a viscosity contrast $\mu_n/\mu_w = 1$ instead of 10 in the base case show that there is no influence of this parameter on the macroscopic relative permeabilities. Finally, the exponents n_f and n_m are changed from 2 to 3 in a few cases; the change of n_f induces negligible changes in the relative permeabilities.

7.6 Comparison with a capillary dominated model

Underlying the macroscopic description of steady two-phase flows in terms of relative permeabilities independent of the driving pressure gradient magnitude is the assumption that the spatial phase distribution is not significantly influenced by the flow, with respect to the equilibrium rest state. This corresponds to small capillary numbers $C \ll 1$ (cf. eqn 7.13).

In this approximation, the local distribution of the phases in the pore volume is assumed to be determined by capillary forces only, and each phase flows through its own system of channels as if the other phase were immobile. Thus, the two flows are actually decoupled, and can be treated as two single-phase flows in a medium where the local permeability is determined by saturation, via the local relative permeabilities.

These considerations, together with the additional feature that the fractures are nearly saturated with the non-wetting fluid, can be applied to devise a simple model for the prediction of the global relative permeabilities.

According to eqn 6.17, the macroscopic single-phase permeability of a fractured medium can be written as

$$K_{\text{eff}} = K_m K'_{\text{eff}}(\rho', \sigma') \tag{7.23}$$

where ρ' is a measure of the network density. In the following, we write in short $K'_{\text{eff}}(\sigma')$ since ρ' is fixed. The function K'_{eff} was the main topic of Chapter 6.

Suppose that in the rest state the mean saturation of the wetting fluid $\overline{S}_w$ corresponds to a capillary pressure $p_{c,r}$. The saturation in the matrix S_{wm} is nearly equal to $\overline{S}_w$, and the associated relative permeabilities are denoted by $K^0_{r,i}$. The saturation in the fractures results from eqns 7.7 and 7.8, and it is generally close to 1, except for very large mean saturations $\overline{S}_w$. The corresponding relative transmissivities in the fractures are $\sigma^0_{r,i}$, with $\sigma^0_{r,w} \ll 1$.

Consider first the wetting-phase flow. The fractures have a negligible contribution to the flow, since $\sigma^0_{r,w} \ll 1$. In the present model, the fractures do not present any resistance to cross-flow, and thus they are neutral with respect to the wetting fluid. Hence, the wetting fluid flows through a uniform medium with apparent permeability $K_m K^0_{r,w}$, and the global wetting phase relative permeability is

$$\overline{K}_{r,w} = \frac{K_m K^0_{r,w}}{K_{\text{eff}}} = \frac{K^0_{r,w}}{K'_{\text{eff}}(\sigma')} \tag{7.24}$$

On the other hand, the non-wetting fluid flows through a fractured porous medium with apparent matrix permeability $K_m K^0_{r,n}$ and fracture transmissivity $\sigma \sigma^0_{r,n}$. Denote by σ'_n the dimensionless ratio defined similarly to σ' for the non-wetting phase

$$\sigma'_n = \frac{\sigma \sigma^0_{r,n}}{L \ K_m K^0_{r,n}} = \sigma' \frac{\sigma^0_{r,n}}{K^0_{r,n}} \tag{7.25}$$

Then, the global apparent permeability of the fractured medium is given by eqn 7.23 as $K_m K^0_{r,n} K'_{\text{eff}}(\sigma'_n)$, and the corresponding relative permeability for the non-wetting phase is

$$\overline{K}_{r,n} = \frac{K_m K^0_{r,n} K'_{\text{eff}}(\sigma'_n)}{K_{\text{eff}}} = K^0_{r,n} \frac{K'_{\text{eff}}(\sigma'_n)}{K'_{\text{eff}}(\sigma')} \tag{7.26}$$

The predictions eqns 7.24 and 7.26 are compared with the full solution of the two-phase flow equations by Bogdanov *et al.* (2003b). For the studied range of parameters and for the two fractured porous media which are studied there, the predictions are always within 20% of the full calculations.

This methodology is illustrated by Exercise 7.2.

Exercises

(7.1) Use the van Genuchten relations with $n_m = n_f = 2$ and $\kappa = 10^{-3/2}$ and determine S_{wf} for $S_{wm} = 0.2$, 0.5 and 0.8.

(7.2) (i) Calculate the relative permeability of the fractured porous medium for the wetting phase in the capillarity dominated model for $S_{wm} = 0.2$, 0.5 and 0.8. Take $\sigma' = 100$. The other conditions are given by Exercise 7.1.

(ii) Same question for the non-wetting phase. Use the approximation based on the Snow formula to estimate the single-phase permeability of the fractured porous medium.

8 Concluding remarks

8.1 Introduction 146
8.2 Numerical 146
8.3 Other phenomena 156
8.4 Where do we stand? 158

8.1 Introduction

The purpose of this chapter is to provide some information which has not been given previously and to complete the overview of our works with some easily understandable material. It is always pleasant to finish a course at a slow pace, and in a relaxed manner!

This final chapter is organized as follows.

Section 8.2 is devoted to the meshers' and solvers' performances.

Section 8.3 lists a number of situations and phenomena related to fractured porous media and which are not addressed in this book.

Section 8.4 concludes the book.

In order to make the comparisons easy, all the CPU times are given for a computer frequency of 3 GHz. However, various parameters such as compiler options and memory cache size can influence the CPU time to a greater degree than clock frequency, in these computations which involve much search in unstructured arrays. Hence, the indications given here are only typical orders of magnitude.

8.2 Numerical

8.2.1 Two-dimensional mesher

Progress in computational resources over the past 15 years plays only a minor role in the ability of Mourzenko *et al.* (2011b) to handle larger samples and particularly much denser networks than Koudina *et al.* (1998). CPU time and memory requirements were not an issue even then. The limitation was in the meshing procedure which was less and less successful as the network density increased. As stated by Koudina *et al.* (1998), 100% success was achieved for $\rho' \leq 6$, whatever the sample size (i.e. the number of fractures), and 90% when $\rho' = 10$. Then the success rate dropped dramatically, to 40% when $\rho' = 12$ and to virtually zero when $\rho' = 16$.

Many improvements have been implemented since then. Without any change in the principle of the method, the algorithm can now cope without fail with the very intricate sets of intersection lines that arise at very large ρ'. However, a new limit is now reached in terms of computation time which becomes very large when $\rho' > 100$. The advancing front technique is appropriate when much of the fracture area is to be triangu-

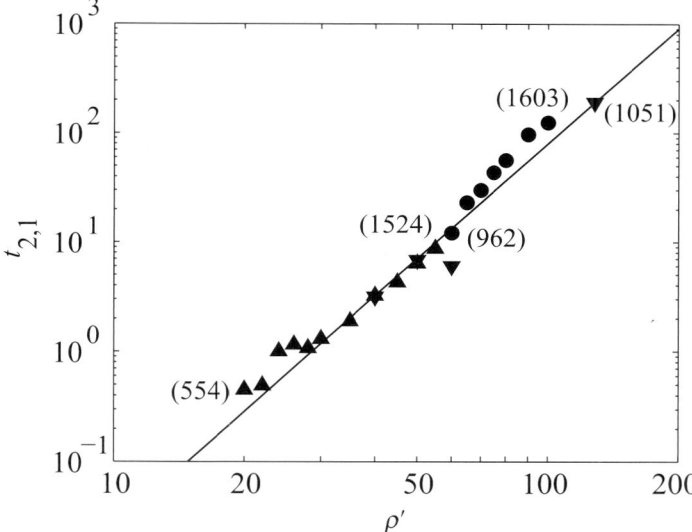

Fig. 8.1 The CPU time $t_{2,1}$ (in seconds) necessary to triangulate one fracture as a function of the network density ρ' for IOUD networks of hexagonal fractures. Data are for $\delta'_M = 1/4$ and a 3 GHz processor: $L/R = 4$ (▼), 5 (●), 6 (▲). The numbers denote the number of fractures in the networks. The solid line is eqn 8.1.

lated between the initial front made of the contours and the intersection lines, but is not when the fractures are already partitioned into many small pieces by a dense network of intersection lines. A different (and probably simpler) approach could be advisable, but it has not been implemented.

An illustration of the numerical effort necessary for triangulating very dense random fracture networks contained in a unit cell of size $L = L'R$ is presented in Fig. 8.1. The CPU time $t_{2,1}$ necessary to triangulate one fracture in a network of dimensionless density ρ' is given by the approximate power law

$$t_{2,1} \approx 8 \cdot 10^{-6} \rho'^{3.5} \text{ seconds} \qquad (8.1)$$

This time is given for the value $\delta'_M = 1/4$ of the discretization parameter that we most commonly use. For finer discretizations, time increases roughly as δ'^{-2}_M. For the sake of illustration, the rightmost point in Fig. 8.1 corresponds to $L' = 4$, $\rho' = 128$, 1051 fractures and the meshing of the whole network requires about two days. Conversely, the leftmost point corresponds to $L' = 6$, $\rho' = 20$, 554 fractures and the whole network is meshed in less than four minutes.

However, one may wonder if calculations for such high densities are necessary. As far as the macroscopic permeability of a fracture network is concerned, one would not learn anything new. Indeed, it was shown in Section 5.4 that the Snow equation is reached and some semi-empirical expressions such as (5.16) were derived in comparison with the numerical results. Of course, new calculations would be necessary for other types of networks, but it is expected that some asymptotic regime should be reached whatever the network characteristics.

8.2.2 Three-dimensional mesher

Performance of the three-dimensional mesher

The performance of the three-dimensional mesher of the porous medium located in between the fractures is shown in Table 8.1. This table analyzes the influence of the two major parameters which are the ratio $L' = L/R$ and the dimensionless density ρ'.

Currently, the range of certain success is about a thousand fractures which correspond to nearly a million tetrahedra.

The memory requirements are not listed in Table 8.1 since they are never a limiting criterion. For instance, meshing a domain of size $L' = 8$ containing a network of 526 fractures ($\rho' = 8$) with $\delta'_M = 1/4$ only requires 170 Mbytes. However, the meshing time t_3 is an important issue. It follows a quadratic law as a function of the numbers N_1 of nodes or N_4 of volume elements

$$t_3 \approx 10^{-5} N_1^2 \quad \approx 4 \cdot 10^{-7} N_4^2 \text{ seconds} \qquad (8.2)$$

Since the numbers of nodes and tetrahedra in the mesh are not always easily foreseeable, in practice, it is more convenient to relate t_3 to the network characteristics. Many tests with various values of all the parameters suggest the empirical formula

$$t_3 \approx 8 \cdot 10^{-7} \left(\frac{L}{\bar{s}}\right)^2 \left(\frac{L}{\delta_M}\right)^6 \text{ seconds} \qquad (8.3)$$

where $\bar{s}$ is the mean spacing of the intersections of a scanline with fractures in the network, given by eqn 3.14e. The factor $(L/\delta_M)^6$ in eqn 8.3

Table 8.1 Mesher performance in 3D. Each case is characterized by four data which are detailed in one array; number of fractures N_0, of nodes N_1, of tetrahedra N_4 and total meshing time t_3.

ρ'	4	8	12	16
$L' = 4$			99 10 600 58 000 14 mn	131 13 600 76 000 39 mn
$L' = 6$		222 25 200 140 000 104 mn	333 32 800 188 000 520 mn	
$L' = 8$	263 42 300 234 000 5 h	526 57 000 328 000 27 h		
$L' = 12$	887 127 000 724 000 90 h			

is the square of the volume to be meshed, expressed in the mesh unit δ_M. The ratio $(L/\overline{s})^2$ traduces the influence of the tesselation of the fracture surface by intersection lines. In view of eqn 3.14e, it introduces a quadratic dependence on the density ρ'.

Recently, the range of addressable situations has been considerably increased.

New or improved ways were devised to handle tricky situations such as nearly parallel objects, warped "plane" objects, nearly coinciding nodes, Schönhart polyhedra, and other difficult cases. These situations are rare but almost certain to occur in very large or dense systems.

Some features have been added to handle T-shaped intersections. Moreover, data from commercial packages such as FRACMAN and AMIRA can be imported.

Many technical details and small improvements were introduced. Their combination allows a replacement of the binary notion of success/failure by a tolerance, in terms of fraction of unmeshed volume (typically, 10^{-6}). Finally, more than 10^4 fractures can be handled, which corresponds to millions of volume elements.

Therefore, the CPU time is certainly the main concern here.

Domain decomposition and parallelization

In order to diminish the CPU time, domain decomposition has been successfully used. For instance, when a new point is inserted in the porous space (cf. Fig. 6.3), several verifications have to be done since the three triangles which are inserted should not cross any existing element. The number of tests is proportional to the number of other elements in the network; therefore, the existence of the quadratic law (8.2) is easily understandable.

One way to break this quadratic rule is to split the domain into smaller pieces. Let us illustrate the methodology by the example displayed in Fig. 8.2. The initial domain is divided into eight blocks by the three mid-planes indicated by the thick black lines. These three mid-planes are considered as fractures and they are meshed together with all the fractures contained in the initial domain. When this triangulation is completed, the volume of each of the eight blocks can be meshed independently of the others.

The computational time for the decomposed domain can be tentatively approximated in the following way. Suppose that a domain of size L is meshed into N_4 volume elements; the time to build this mesh is proportional to N_4^2. If the domain is decomposed into N_b blocks of size L_b, the mesh of each block contains N_4/N_b tetrahedra and its construction only requires a time which is proportional to $(N_4/N_b)^2$. Since there are N_b such blocks, the domain decomposition is expected to divide the total meshing time by a factor N_b.

Of course, things are not so simple since the decomposition into blocks and the final assembling of the mesh pieces introduce a significant time overhead. In addition, the decomposition in itself slightly increases the final number of volume elements.

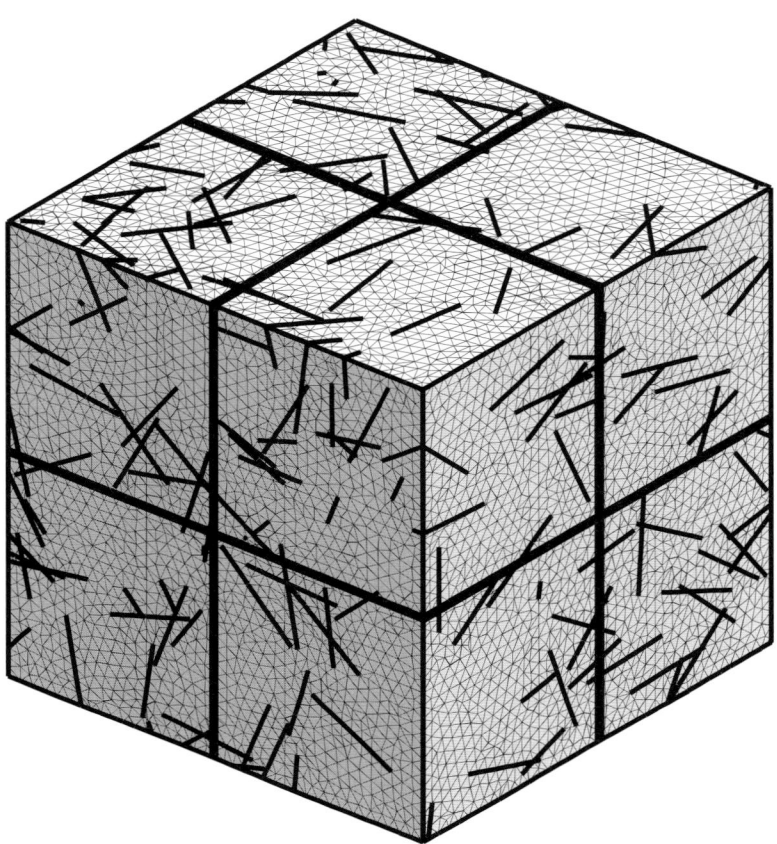

Fig. 8.2 Domain decomposition. The initial domain is decomposed into eight blocks by the three mid-planes.

Systematic tests are summarized in Fig. 8.3, for a wide range of the parameters with $L' =$ 6, 8, 12 and 16, $\rho' =$ 2, 4 and 8, $\delta'_M = 1/4$ and $1/8$ and a decomposition into $N_b =$ 1, 8, 27, 64, 216 and 512 blocks. The total meshing time t_3 is plotted in Fig. 8.3a as a function of the block size in mesh unit L_b/δ_M. The data are very scattered but the lines which connect points corresponding to identical parameters except for the block decomposition show that in all cases t_3 first decreases when the block size decreases and then increases again. This is even more visible in Fig. 8.3b where t_3 is normalized by N_4^2. A clear optimum exists when $L_b/\delta_M \approx 10$, and t_3 is then about $10^{-8} N_4^2$. This provides a very significant improvement upon eqn 8.2 and a simple practical guideline for the optimal setting of the block size.

The computation time $t_{3,b}$ for the meshing of one block is shown in Fig. 8.3c, normalized by the squared relative block size $(L_b/\delta_M)^6$. The data are very scattered, but the detrimental effect of too small a block size is clearly visible. A further normalization by the ratio $(L/\overline{s})^2$ is fairly successful in gathering the results in the range $L_b/\delta_M \geq 10$, which can be approximately represented by (Fig. 8.3d).

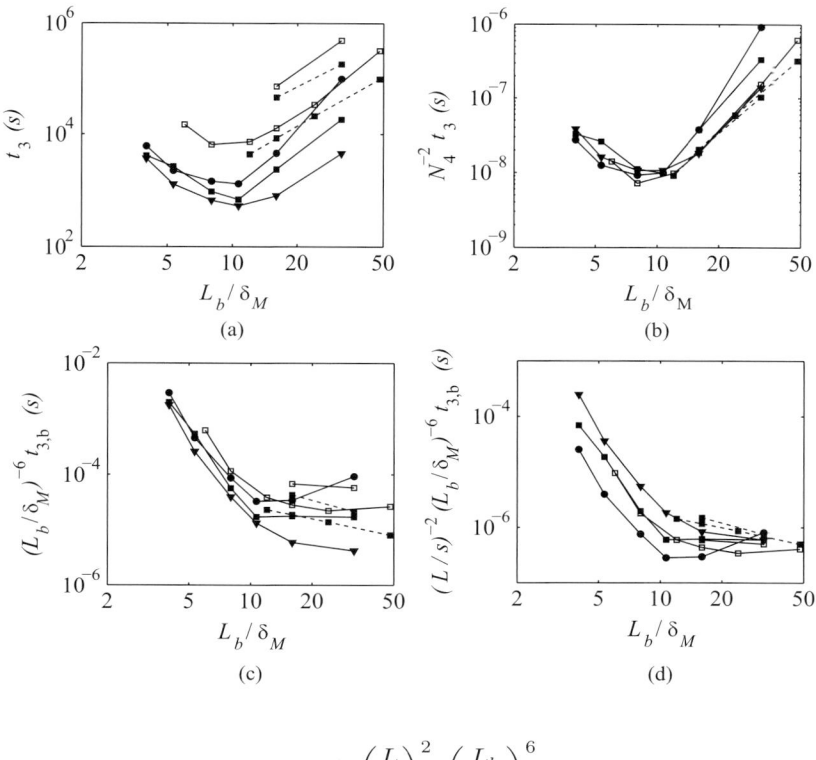

Fig. 8.3 Meshing time in seconds as a function of the block size L_b/δ_M. (a) total time t_3; (b) normalized total time $N_4^{-2} t_3$; (c); normalized time $(L_b/\delta_M)^{-6} t_{3,1}$ for the meshing of one block; (d); normalized time $(L/\bar{s})^{-2}(L_b/\delta_M)^{-6} t_{3,1}$ for the meshing of one block. Data are for $\rho' = 2$ (▼), 4 (■) and 8 (●); $\delta'_M = 1/4$ (solid lines) and $1/8$ (broken lines); $L' = 6$ or 8 (solid symbols) and 12 or 16 (open symbols). The lines join data for identical parameters except for the block decomposition.

$$t_{3,b} \approx 8 \cdot 10^{-7} \left(\frac{L}{\bar{s}}\right)^2 \left(\frac{L_b}{\delta_M}\right)^6 \text{ seconds} \quad (8.4\text{a})$$

The computation time for the whole domain is obtained by multiplying $t_{3,b}$ by the number $(L/L_b)^3$ of blocks,

$$t_3 \approx 8 \cdot 10^{-7} \frac{L^5 L_b^3}{\bar{s}^2 \delta_M^6} \text{ seconds} \quad (8.4\text{b})$$

Of course, eqn 8.4 reduces to eqn 8.3 when no decomposition is applied, i.e. $L_b = L$.

Domain decomposition is ideally suited to parallel computation. Such a scheme, which was actually not implemented in our group, is displayed in Fig. 8.4. Calculations could be done on a computer with distributed memory, or even on physically distinct computers.

Some recent examples

Recent examples which have been addressed are gathered in Fig 8.5. They can be commented on as follows.

Cases (a) and (b) are detailed by Mourzenko *et al.* (2011c) who addressed the simulation of single-phase transient compressible flows and well tests. (a) represents the modelization of an experimental field in Poitiers, France. It is composed of three families of subvertical fractures and of one subhorizontal one; the total number of fractures is

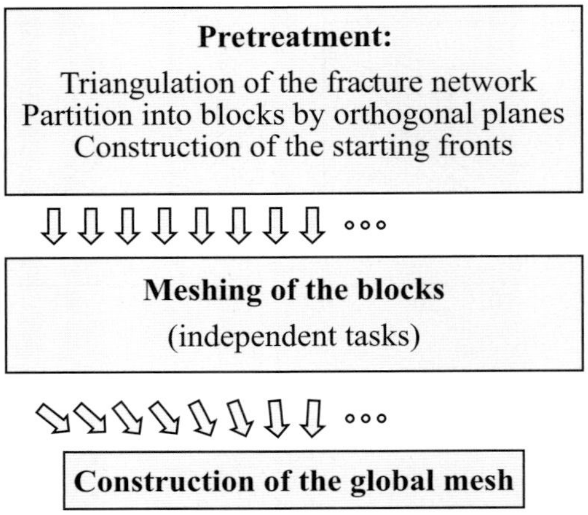

Fig. 8.4 Straightforward parallelization associated with domain decompositions such as the one displayed in Fig. 8.2.

```
                    ┌─────────────────────────────────────────┐
                    │              Pretreatment:              │
                    │ Triangulation of the fracture network   │
                    │ Partition into blocks by orthogonal planes │
                    │  Construction of the starting fronts    │
                    └─────────────────────────────────────────┘
                       ⇓ ⇓ ⇓ ⇓ ⇓ ⇓ ⇓ ⇓  ooo
                    ┌─────────────────────────────────────────┐
                    │          Meshing of the blocks          │
                    │          (independent tasks)            │
                    └─────────────────────────────────────────┘
                       ⇘ ⇘ ⇘ ⇘ ⇓ ⇓ ⇓  ooo
                    ┌─────────────────────────────────────────┐
                    │      Construction of the global mesh    │
                    └─────────────────────────────────────────┘
```

equal to 2466; the porous medium is meshed by $3.3 \cdot 10^6$ tetrahedra. Since well tests are simulated, an additional porous medium has been added around the fractured porous medium; its properties are equal to the macroscopic permeability of the fractured medium. In such a way, the reflection of the pressure signals by the overall boundaries does not disturb the experiments.

(b) corresponds to an oil field with seven deterministic large accidents or faults and 719 randomly generated small fractures. The total number of tetrahedra is equal to $3.5 \cdot 10^5$.

(c) is taken from Thovert *et al.* (2011) who analyzed the excavation damaged zone (EDZ) around a tunnel dug in a clay formation. The fracture network is composed of large pre-existing faults and of the small fractures of the EDZ, in total 9300 fractures and $1.05 \cdot 10^6$ tetrahedra.

(d) corresponds to a fractured porous medium with power-law size distribution of fractures modeled by hexagons. There are 9650 fractures separated by $1.1 \cdot 10^6$ tetrahedra.

A remarkable feature of the volume mesher is that it is not limited to fracture networks. It may start from any set of triangulated surfaces and therefore it may be used to mesh porous media. Two such examples are given in Fig. 8.6. Malinouskaya *et al.* (2008) used unstructured meshes made of tetrahedra in order to discretize and solve various equations such as the Stokes equations on the pore scale in the grain packing shown in (a); note that the pore and the solid spaces are meshed starting from the same triangles located at the solid-pore interface. Gerbaux *et al.* (2010) meshed metallic foams starting by the triangulation of the metal–pore interface provided by the software AMIRA since the foam geometry was measured by computed micro tomography (see b); the Stokes equation was solved and the foam permeability determined.

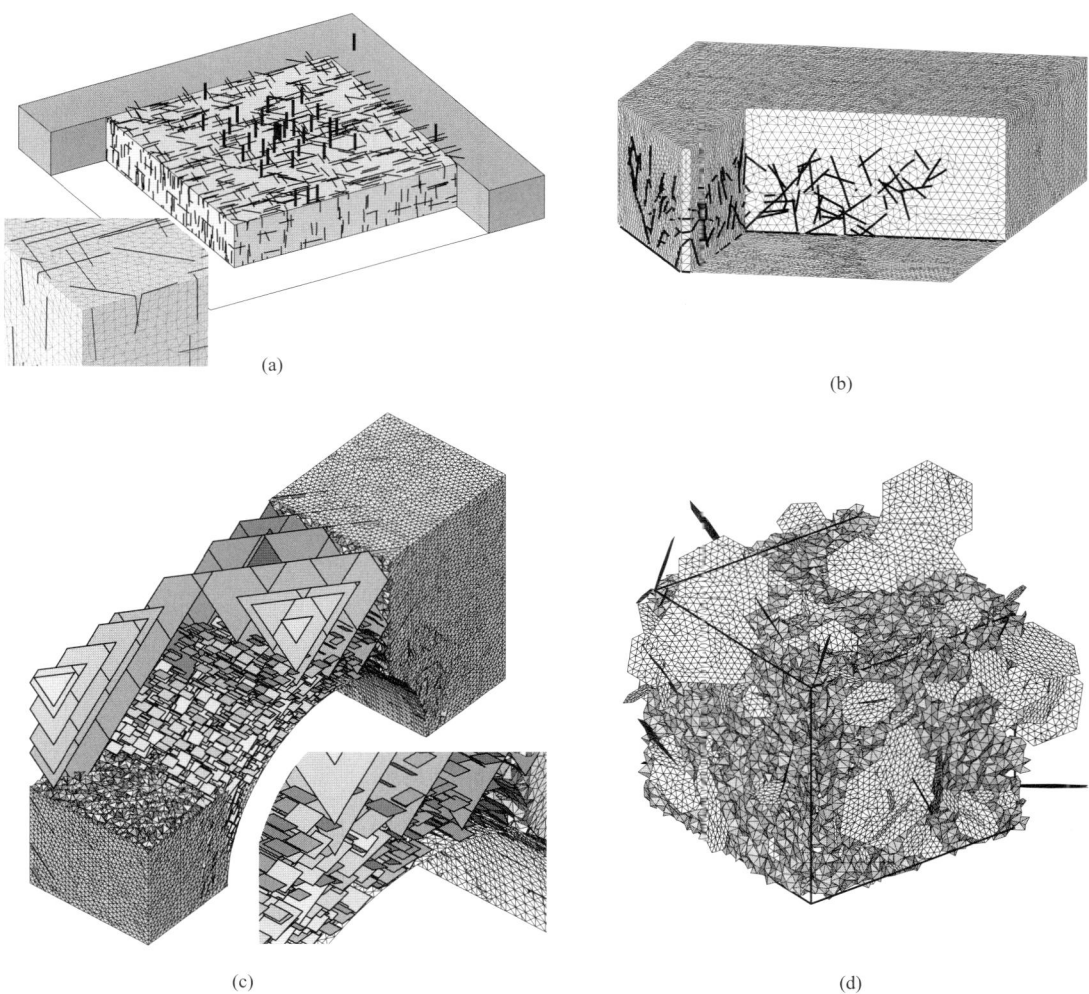

Fig. 8.5 A gallery of recent examples. (a) Experimental field in Poitiers (SEH) modeled by four families of fractures. (b) Oil reservoir built from deterministic faults and random fractures. (c) Fractured porous medium around a gallery with large pre-existing faults and small EDZ fractures. (d) Simulated fractured medium with power-law size distribution.

8.2.3 Single-phase flow

The flow solver performances can be quantified in various ways. The simplest one is the time requirement t_f for a steady flow solution. It corresponds to the determination of the pressures at the mesh nodes which satisfy the Darcy equation under stationary boundary conditions. Recall that this is done using an iterative method, based on a conjugate gradient algorithm, starting from an arbitrary initial guess. This is the typical situation when determining an effective permeability tensor. It can also correspond to one time step in the simulation of a transient flow situation, if the temporal steps are long and the pressure field is

154 *Concluding remarks*

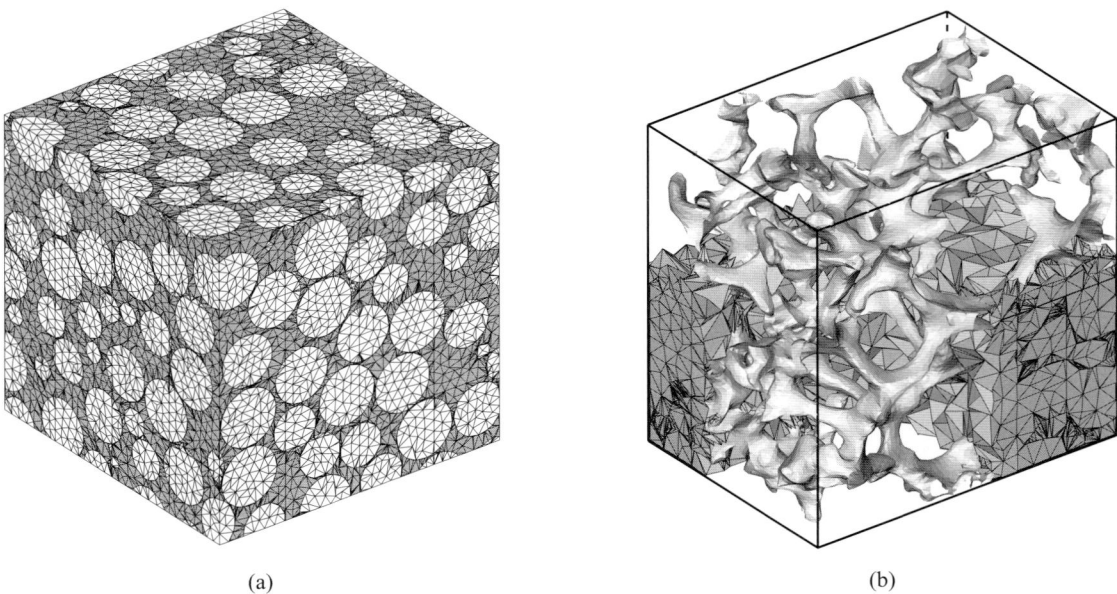

Fig. 8.6 A gallery of recent examples. (a) Grain packing. (b) Foam.

significantly modified. On the other hand, in simulations of transient phenomena with a fine time resolution, the pressure field is only slightly modified over each time step and only a few elementary iterations of the solution algorithm are necessary. Hence, knowledge of the CPU time $t_{f,1}$ for this elementary operation is also of interest.

Illustrative CPU times relative to the recent applications displayed in Figs 8.5a and b are given in Table 8.2.

Note first that the memory size is directly proportional to N_1,

$$M \approx 3\text{kB per node} \tag{8.5}$$

The elementary time $t_{f,1}$ only depends on the number of elements in the mesh, and since the numbers of triangles and tetrahedra are always of the order of 12 N_1 and 6 N_1, respectively, $t_{f,1}$ should be approximately proportional to the number of nodes. However, the data in Table 8.2 have been obtained with processors of different types, which probably explains some variations. A typical estimate can be given as

$$t_{f,1} \approx 2 \cdot 10^{-6} \, N_1 \text{ seconds} \tag{8.6}$$

On the other hand, the number N_{it} of iterations required to reach a solution depends on the conditioning of the equations and on the prescribed tolerance, in addition to the mesh size. Tolerance used for the data in Table 8.2 is set in order to predict the effective permeability with a relative accuracy of less than 1%. A better convergence may be necessary

Table 8.2 CPU times on a 3 GHz processor for the resolution of the Darcy equation. The oil reservoir and the Poitiers field are displayed in Figs 8.5b and a, respectively.

Simulations	Oil reservoir	Poitiers	Poitiers	Poitiers	Poitiers
Nodes	64 000	196 000	564 000	564 000	564 000
Memory (MB)	184	~ 600	$\sim 1\,500$	$\sim 1\,500$	$\sim 1\,500$
σ'	~ 1	$\sim 10^3$	~ 10	$\sim 10^2$	$\sim 10^3$
CPU time $t_{f,1}$ (sec) (1 elementary iteration)	0.23	0.49	0.82	0.68	0.81
Number N_{it} of iterations (1 flow solution)	800	1350	640	1020	1310
CPU time t_f (min) (1 flow solution)	~ 3	~ 11	9	11.5	18
CPU time (h) (full reservoir history, ~ 3000 time steps)	~ 15				

if the calculated flow field is to be subsequently used for the simulation of solute transport problem. Then, N_{it} increases as a logarithmic function of the convergence criterion. Poor conditioning of the system of equations can result from the large contrast in the hydraulic properties of the matrix and fractures. Thus, N_{it} increases with the dimensionless transmissivity σ'. Geometrical features of the fracture network can also spoil the conditioning, e.g., when it is poorly connected and concentration of flow occurs in critical sections. Finally, N_{it} also increases with the problem size, roughly as the square root of N_1.

Let us finally mention that 80–95% of the CPU time is spent in a few very localized spots, and more so as the problem size increases. These time intensive steps are loops over the control volumes contained in matrix/vector product subroutines. This is an ideal situation for a parallel execution.

8.2.4 Two-phase flow

It is impossible to give general a priori estimates of the computation time for two-phase flow simulations, even very approximately, since it depends not only on the mesh characteristics and matrix/fracture properties, but also on the mean saturation and on the non-linear constitutive laws for the relative permeabilities and capillary pressure as functions of the saturation.

The best that we can do is to provide illustrative indications, based on the simulations described in Figs 7.5 and 7.6a for a fractured medium of the type shown in Figs 7.4a,b. The mesh contains $N_1 \approx 1800$ nodes and $N_4 \approx 11\,000$ tetrahedra. The memory requirements are about ten times greater than eqn 8.5 for single-phase flow, since two pressure fields need to be determined and many more physical coefficients to be stored on a per element basis.

Each of the simulations corresponding to the three columns in Fig. 7.5, starting from different initial distributions of the fluids required about 10 min. However, this time is strongly dependent on the mean saturation $\overline{S}_w$. The computations for each flow solution conducted for the determination of the relative permeabilities shown in Fig. 7.6a took about 5 min for $\overline{S}_w \approx 0.10$, about 10 min in the intermediate range $\overline{S}_w \sim 0.50$, and nearly 3 hours when $\overline{S}_w \approx 0.90$. The reason is that S_w can become very close to one in the last case in some regions upstream of the fractures. The derivative $\partial p_c / \partial S_w$ diverges in this limit which makes convergence of the Picard scheme very difficult.

This effect, and more generally the saturation variations induced by the flow and illustrated in a mild situation in Fig. 7.5, increase with mean saturation, fracture transmissivity and global flow rate.

8.2.5 Conclusions

Although the typical computation times for the successive operations of 2D meshing, 3D meshing and flow solution are not always very accurately predictable, they are sufficiently contrasted to clearly identify the limiting step.

The 2D meshing of a network containing about 10^3 fractures is a matter of seconds, unless the dimensionless density ρ' is very large. The last line in Table 8.1 shows that direct 3D meshing of the matrix space can take several days. Finally, the data in Table 8.2 show that a single-phase flow solution requires only a few minutes, even in the SEH which contains significantly more fractures and mesh elements.

Hence, 3D meshing is clearly the bottleneck, and optimization of the performances should focus primarily on this aspect. The domain decomposition is a great step in this direction since it allows one to build the 3D mesh in hours instead of days. However, Fig. 8.3 shows that there is a limit to this approach, since excessive decomposition has a detrimental effect on the computation time.

8.3 Other phenomena

Some phenomena were addressed with the same general methodology, but were not studied—and often even not mentioned—in this course though they may be of general interest. They can be classified into three groups.

Several phenomena were studied in single fractures. Mourzenko *et al.* (1996b) studied deposition of a single solute in a single fracture in the limit where the geometrical changes are very slow compared to the average fluid velocity; examples of deterministic fractures, random but uncorrelated fractures, Gaussian and self-affine fractures were studied for four different values of the Péclet and the Péclet-Damköhler numbers. This study was extended by Békri *et al.* (1997) to successive cycles of deposition-dissolution for the same four values of these numbers; some conditions yield reversible cycles and some others an almost fixed configuration. The mechanical deformation of a single fracture and its influence on fracture permeability were addressed by Mourzenko *et al.* (1997); a mean field approximation was derived that may include non-linear effects due to variations in the contact surface. Finally, electroosmotic phenomena in single fractures were studied by Marino *et al.* (2000); all the electroosmotic coupling coefficients could be gathered into a single relationship which depends on a single characteristic length scale applicable to every configuration.

This book covers almost all our work on fracture networks, with two exceptions. The first is the analysis of traces in cylindrical tunnels. The intersections between three-dimensional fracture networks consisting of constant and variable size disks with a gallery were systematically analyzed by Gupta and Adler (2006). Exact and approximate analytical formulae were derived for the numbers of partial or full intersections. These predictions were verified by Monte Carlo simulations. Trace length distributions were studied using numerical techniques for various disk distributions. These simulations showed that differences between various disk distributions are very small when trace length distributions with the same average and standard deviations are compared. The second aspect which is not covered here is the microscale study of solute transport at the intersection of fractures by Mourzenko *et al.* (2002). The validity of the classical models of solute mixing, namely stream tube routing and perfect mixing, generally used in large-scale modeling of solute transport in fracture networks, was analyzed by comparing their predictions with the results of direct Lagrangian numerical simulations.

Patriarche *et al.* (2007) conducted a study along the 128 m long Roselend tunnel (French Alps) which is briefly mentioned in Section 3.3.4. Fractures can be classified into two families: large fractures which intersect the tunnel and small fractures which partially intersect it. Three different zones in the tunnel are distinguished with mild, low and high water fluxes, starting from the entrance. A stereological analysis of the trace length probability densities of small fractures provides the fracture diameter probability density distribution which is best approximated by a power law. Large fractures are assumed monodisperse, with a 5 m estimated radius. The generated fracture networks obtained by combining large and small fractures do percolate, while networks consisting of only small fractures do not percolate. Computed macroscopic permeabilities of the fracture networks are in agreement with the observed water fluxes when the fracture permeability is a power law of its lateral extent with an exponent equal to 3.

Finally, electroosmotic phenomena in fractured porous media were also briefly addressed approximately by Adler (2001).

8.4 Where do we stand?

We have certainly made progress towards the general objective of this book which is to estimate the macroscopic properties of fractures, fracture networks and fractured porous media from easily measurable quantities.

A general methodology has been implemented in order to calculate the macroscopic properties of these three possible media. Concerning fracture networks and especially fractured porous media, the difficulty lies in the meshing and not the resolution of the equations themselves, at least not of the elliptic equations which govern flow on the Darcy scale.

The numerical results which have been systematically derived for generic structures such as I^2OUD networks, networks with power-law size distributions and anisotropic networks, could be rationalized by the systematic use of the excluded volume. The dimensionless density of fractures ρ' equal to the number of fractures per excluded volume yields master curves on which the data relative to quite different structures can be gathered. In many cases, the macroscopic properties of the media only depend on ρ' and not on the shape of the fractures.

Stereological relations have also been derived and they enable us to derive, with some hypotheses, the dimensionless density. And the macroscopic properties follow on.

The efficiency of this methodology has been fully verified in a few cases only (Gonzalez-Garcia *et al.*, 2000; Patriarche *et al.*, 2007).

What is really missing at the moment are experimental results on real sites where the fractures have been recorded and characterized, and where overall flow measurements have been performed. A good example, but not the perfect one obviously is the work done by Patriarche *et al.* (2007).

Such examples should be multiplied in order to place the theoretical results on a firm experimental basis.

Notation

Unless otherwise stated, numbers in parentheses after descriptions refer to equations in which symbols are first used or thoroughly defined; when this is not possible, the section where they are defined is indicated. Boldface symbols are vectors or tensors. Primed quantities are dimensionless. Symbols with several meanings throughout the book are listed under the name Generic; symbols that appear infrequently or in one section only are not listed; the precise meaning of these last two types of symbols is given in the section or the subsection where they appear.

Chapter 7 has separate nomenclature, but also contains symbols which appear in the general list.

Generic: letters with multiple meanings

Latin letters

d	space dimension
$f(\cdot)$	function (1.4)
i, j, k, l, m, n	integers. i may also be the complex number $i^2 = -1$ (Section 2.6.1)
ℓ	characteristic length scale (Section 1.2)
L	length of a fracture or other characteristic length (Section 1.2)
S	surface, cross-sectional area (Section 1.3)
$\boldsymbol{x}$	2 or 3D position vector

Greek letters

α, β	coefficients
ϵ	generally a small parameter, or possibly a porosity
θ	angle (dip, polar coordinate)
ϕ	angle (azimuth, polar coordinate)

General nomenclature

Mathematical symbols

Averages

$\overline{(.)}$	spatial average over a line, a surface or a volume (Sections 2.2.1 and 2.6.2)
$\overline{\overline{(.)}}$	two-stage spatial average

$\langle . \rangle$ statistical average (Sections 2.2.1 and 2.6.2)

$\langle \overline{X} \rangle_Y$ conditional averages of $\overline{X}$ over domains which share a common value y of some random variable Y (2.46)

Operators

$\| \ \|$ standard Cartesian norm (5.13)

$d\ell$ line differential element (4.12)

ds differential surface element

∇ three-dimensional gradient operator

∇_s gradient operator; it works in the fracture plane only where the field to which it is applied, is defined; see eqn 5.1 and Exercise 5.1

Others

erfc   complementary error function (Gradshteyn and Ryzhik, 1965)

i complex number: $i^2 = -1$ (Section 2.6.1)

n-rectangle   rectangle with an aspect ratio equal to n

Prob(ω) probability of the event ω (2.34)

$\hat{X}_{st}$ Fourier transform of the field X_{mn} (2.30)

$\check{Y}$ spatially periodic function (2.25a)

Acronyms

I^{2}OUD or IIOUD   isotropically oriented and uniformly distributed network of identical fractures (Section 3.4.1); this corresponds to the reference case which is denoted by the subscript r

IOUD   isotropically oriented and uniformly distributed (Section 3.4.1); the fractures may be of different sizes or different shapes

Subscripts

L quantity obtained by solving the Laplace equation (Section 4.3.1)

r reference case, i.e. relative to isotropically oriented and uniformly distributed fractures (Section 3.4.1)

R quantity obtained by solving the Reynolds equation (Sections 4.2.2 and 4.3.2)

S quantity obtained by solving the Stokes equation: Chapter 4 only

S Snow approximation: Chapters 5, 6, 7

0 indicates a characteristic value which is defined locally

Superscripts

$'$ a prime denotes a dimensionless quantity; the most important is the dimensionless fracture density ρ' defined by (3.6)

Latin letters

a	size of the elementary cube in Chapter 4
a	exponent of the power law (3.26) in Chapters 3, 5, 6, 7
$a(m)$	coefficients of the linear combination (2.14)
A	surface of the fracture projected onto the xy-plane (Section 2.4.1)
A_{pc}	percolating connected component (Section 2.4.3)
A_{npc}	non-percolating connected component (Section 2.4.3)
$b(\boldsymbol{x})$	local aperture of the fracture at point $\boldsymbol{x}$ (Section 1.2, eqns 2.2, 2.21)
b_m	separation (or distance) between the two average planes of a fracture (2.4)
b_S, b_R	equivalent mean apertures for permeability (4.17)
b_{cL}, b_{cR}	equivalent mean apertures for conductivity (4.31)
b_0	characteristic fracture aperture (4.41)
c	chord length (Section 3.2); otherwise, c solute volumetric concentration in Sections 4.3 and 4.4
C_g	autocorrelation function of the function g (2.5b)
$\mathcal{C}$	total trace length per unit surface (3.21)
$f_i(\theta, \phi)$	isotropic probability density function (3.36)
$f_F(\theta, \phi)$	Fisher probability density function (3.37)
$g(c)$	probability density of the chords (Section 3.2)
$h^{\pm}(\boldsymbol{x})$	fluctuations of the surfaces S_p^+ and S_p^- around the average planes (2.1)
$h_0^{\pm}$	average elevations of the surfaces S_p^+ and S_p^- limiting the fracture (2.1)
H	Hurst exponent (2.10)
I_h	intercorrelation of functions h (2.7a)
$\boldsymbol{I}$	unit tensor
$\boldsymbol{J}$	flow rate per unit width (4.4)
$\boldsymbol{J}_c$	solute flux per unit width (4.26)
$\boldsymbol{J}_R$	local solute flux per unit width in the Reynolds approximation (4.28)
k_S	dimensionless coefficient defined by (5.11d)
$\boldsymbol{k}$	reciprocal vector (2.32b)
K	scalar permeability (1.1e)
K_{eff}	macroscopic permeability of a fractured porous medium (6.9)
$K_{\text{eff},r}$	macroscopic permeability of an IOUD fractured porous medium (Section 6.4)
$K_{\text{eff}S}$	macroscopic permeability of a fractured porous medium by using the Snow approximation (6.12a)
K_f	permeability of a porous material filling the fracture (6.6a)
K_m	bulk permeability of a porous medium (6.1)
K_{nr}	permeability of a fracture network of isotropically oriented and uniformly distributed (I^{2}OUD) identical fractures (Section 5.4)
K_{nh}	dimensional permeability of a network of heterogeneous fractures; K'_{nh} its dimensionless value; Section 5.5.3

$\boldsymbol{K}$	permeability tensor
$\boldsymbol{K}_{nS}$	Snow formula for an anisotropic network of infinite fractures (5.9b)
$\boldsymbol{K}_{nSr}$	Snow formula for an isotropic network of infinite fractures (5.10)
K'_{eff}	macroscopic dimensionless permeability of a fractured porous medium (6.10)
$K'_{\text{eff},r}$	dimensionless macroscopic permeability of an IOUD fractured porous medium $= K_{\text{eff}}/K_m$ (Section 6.4)
$K'_{\text{eff}S}$	macroscopic permeability of a fractured porous medium by using the Snow approximation $= K_{\text{eff}S}/K_m$
$K'_2(\rho'_3)$	dimensionless function (5.27)
$K''_2(\rho'_3, \sigma')$	permeability coefficient of polydisperse networks (6.21a)
$\ell_c, \ell_{c1}, \ell_{c2}$	characteristic correlation lengths (Section 2.2.2)
ℓ_d	characteristic decay length (3.2)
ℓ_p	characteristic dimension of the pores in a porous medium (Section 1.2)
$n(R)\mathrm{d}R$	probability of fracture radii in the range $[R, R+\mathrm{d}R]$ (3.26)
$n_I(\boldsymbol{p})$	number of intersections between a fracture and a line parallel to $\boldsymbol{p}$ per unit length of the line (Section 3.2)
$\boldsymbol{n}$	unit normal to a fracture (Section 3.2)
N_I	number of intersections of a line of length L with traces (Section 3.5)
p	pressure (Section 1.3)
$\boldsymbol{p}$	unit vector parallel to a given line (Section 3.2)
$\boldsymbol{p}_F$	unit vector parallel to the Fisher pole (Section 3.3.3)
P	probability of existence of a bond (Section 2.5.1)
P_c	percolation threshold (Section 2.5.1)
Pe	Péclet number (4.41)
Q	flow rate
Q_c	overall solute flux (4.25)
R	radius of the circle circumscribed to a polyogonal fracture; more generally, half of the lateral extension of a fracture (Section 3.1)
Re	Reynolds number (1.3a)
R_m and R_M	minimal and maximal radii of polydisperse fractures (Section 3.8.1)
$R'_m = R_m/R_M$	(Section 3.8.1)
s_j	spacing number j between two intersections along a line (Section 3.2)
$\overline{s}$	mean spacing between two intersections along a line (3.14e)
S_c	proportion of the fracture surface A where the two blocks are in contact (2.22a)

S_p^+, S_p^-	upper and lower solid surfaces limiting the void space of a fracture (Section 1.2)
S_0	fractional open area where the two blocks are not in contact (2.22c)
$S_{0,\lambda} = 1 - \overline{Z_c(\boldsymbol{x})}$	fractional open area measured over a domain of size λ (Section 2.6.2)
$\mathcal{S}$	total fracture surface per unit volume (5.10); $\mathcal{S}'$ its dimensionless value defined by (6.21b)
$\mathcal{S}(\boldsymbol{k})$	(with various indices) power spectral density which is a function of the reciprocal vector $\boldsymbol{k}$ (or of its modulus k); (2.30b), (2.36); only in Section 2.6
$\mathcal{S}(\boldsymbol{n}_j)$	surface area per unit volume of the family of fractures S_j (5.9b)
t	universal exponent for conductivity (2.27)
$\boldsymbol{u}$	2D lag in the autocorrelation function (2.5b)
$\boldsymbol{v}$	local fluid velocity; v local fluid scalar velocity (Section 1.3)
$\overline{\boldsymbol{v}}$	local seepage velocity in a porous medium (6.1)
$\overline{\boldsymbol{v}}^*$	mean interstitial velocity (4.39b)
$\overline{\boldsymbol{v}}_\perp$	seepage velocity normal to the fracture with a pressure drop Δp (6.4)
$\overline{\boldsymbol{v}}^+$	seepage velocity in the matrix on the side of the unit normal $\boldsymbol{n}$ to the fracture; $\overline{\boldsymbol{v}}^-$ seepage velocity on the opposite side (6.7)
$\overline{\overline{\boldsymbol{v}}}$	overall seepage velocity in a fractured porous medium (6.9), $\overline{\overline{v}}$ its modulus (1.1c)
v_{ex}	dimensionless excluded volume for polydisperse networks (3.32b)
V_{ex}	excluded volume (Section 3.4.1)
$V_{ex,r}$	excluded volume for the reference case, i.e. for isotropically oriented and uniformly distributed fractures (Section 3.4.1)
W	fracture width (Section 1.2)
x, y, z	orthonormal system of coordinates (Section 1.2)
X, Y	one-dimensional random fields (2.13), (2.14)
$\boldsymbol{X}_{n,m}$	spatial period (2.25b)
$z^\pm(\boldsymbol{x})$	elevations of the surfaces S_p^+ and S_p^- limiting the fracture (2.1)
Z_c	phase function of the contact zone (2.22b)

Greek letters

α_K, β_K	coefficients in (5.16b)
β_σ	exponent of the power law for the fracture permeability defined by (5.26)
β_{12}	angle between the normals $\boldsymbol{n}_1$ and $\boldsymbol{n}_2$ to the two fractures 1 and 2 (Section 3.7.1)

$\gamma_s(n)$	semivariogram with a lag n (3.1)
δ	local discretization length (Section 5.3.1); $\delta \leq \delta_M$
δ_M	maximal discretization length (Section 5.3.1)
δ'_M	$= \delta_M/R$
$\Delta\rho' = \rho' - \rho'_c$	(5.15) or $\rho'_3 - \rho'_{3c}$ (6.22)
ζ_i	(i=1,2,3) consistency coefficients which should be equal to 1 (3.25)
η_p	shape factor (3.11)
θ_I	intercorrelation coefficient (2.7b)
κ	coefficient of the Fisher distribution (3.37)
$\langle \Lambda \rangle$	statistical average of the macroscopic conductivity (2.27)
Λ_{co}	upper cut-off for the self-affinity of the surface profiles (2.43b)
$\mathbf{\Lambda}$	conductivity tensor (4.26b)
$\mathbf{\Lambda}_L$	conductivity tensor determined by solving the Laplace equation (4.26b)
$\mathbf{\Lambda}_P$	conductivity tensor for the Poiseuille configuration (4.27c)
$\mathbf{\Lambda}_R$	conductivity tensor determined by solving the Reynolds equation (4.29a)
$\overline{\Lambda}'$	dimensionless conductivity (4.32)
μ	fluid viscosity
ν	universal exponent for cluster size (2.26)
$\boldsymbol{\nu}_\Pi$	unit normal to the observation plane Π (Section 3.7.1)
ξ	average distance between two sites which belong to the same cluster (2.26)
Π_f	probability of percolation of a fracture (Fig. 2.11)
Π	plane; observation plane (Section 3.2.1)
ρ	fracture density defined as the number of fractures per unit volume (Section 3.1)
ρ_b	solid block density (Section 3.4.4)
ρ_c	fracture density at the percolation threshold (Section 3.4.3)
$\rho_{c,r}$	fracture density at the percolation threshold for isotropically oriented and uniformly distributed fractures (3.11)
ρ_f	fluid density
ρ_o	density at the wall (3.2)
ρ'	dimensionless fracture density (3.6)
ρ'_c	dimensionless percolation threshold (Section 3.4.3)
ρ'_3	dimensionless density extended to polydisperse networks (3.32a)
ρ'_{3c}	dimensionless percolation density extended to polydisperse networks (3.34)
σ	fracture transmissivity (4.4)
σ_{av}	unit for fracture transmissivity (4.18)
$\sigma_h, \sigma_{h^+}, \sigma_{h^-}$	surface roughnesses (2.5a)

σ_P	fracture transmissivity for a Poiseuille flow (4.9)
σ_R	fracture transmissivity calculated with the Reynolds equation (4.16)
σ_S	fracture transmissivity calculated with the Stokes equation (4.4c)
Σ_b^2, Σ_h^2	expectations of $\overline{\sigma_b}^2$ and $\overline{\sigma_h}^2$ (2.48)
Σ_p	number of intersections between traces per unit surface (Section 3.2)
Σ_t	surface density of traces or equivalently the number of traces per unit surface (Section 3.2)
σ'_S, σ'_R	dimensionless transmissivities (4.18)
$\varphi(h^\pm)$	probability density of the fluctuations $h^\pm$ (2.5a)
Φ	correction factor for the excluded volume relative to the reference case (3.38)
$\psi_\ell, \psi_t, \psi_c, \psi_p$	correction factors for various quantities relative to the reference case (3.38)
$\boldsymbol{\psi}_K$	tensorial correction factor defined by (5.31) with components $\psi_\perp$ and $\psi_\parallel$
ω	in plane orientation of a fracture (Fig. 3.4 and Section 3.2)
ω	normal resistance of the fracture (6.4)
ω'	its dimensionless value (6.6c)

Specific nomenclature for Chapter 7

Subscripts

$i = w, n$	subscripts which refer to the wetting and non-wetting fluids, respectively
$j = m, f$	subscripts which refer to the porous matrix and the fractures, respectively

Latin letters

g	gravity
$\boldsymbol{J}_i$	flow rates per unit fracture width
$K_{r,i}$	($i = w, n$) relative permeabilities of the porous medium
$\overline{K}_{r,i}$	relative macroscopic permeabilities (7.16), (7.20)
n	exponent for the capillary pressure
p_c	capillary pressure (7.6)
p_i	pressure for fluid i
p_0	characteristic pressure (7.7)
S_{ij}	($i = w, n$) local saturations in the porous medium $j = m$ and in the fractures $j = f$
$\overline{S}_w$	mean global saturation

$\tilde{S}_w$	effective saturation (7.10)
S_{wmr} and S_{wms} (S_{wfr} and S_{wfs}, resp.)	irreducible and maximal saturations in the porous matrix (the fractures, resp.)
t'	dimensionless time defined by (7.18) or by (7.19)
$\overline{v_i}$	local seepage velocity of phase $i = w, n$
V	total pore volume of the reservoir
z	vertical axis oriented downwards

Greek letters

γ	surface tension
$\Delta\rho = \rho_n - \rho_w$	(7.6)
ϵ_m	porosity of the porous medium
ϵ_f	porosity of the fracture
κ	defined by $p_{0,f} = \kappa p_{0,m}$ (7.8)
μ_i	viscosity of fluid i
ρ_i	density of fluid i
$\sigma_{r,i}$ ($i = w, n$)	relative fracture transmissivities
Φ_i	the potential $p_i - \rho_i g z$

References

Adler, P.M. (1992). *Porous Media: Geometry and Transports*. Butterworth-Heinemann, Stoneham, MA.

Adler, P.M. (2001). Macroscopic electroosmotic coupling coefficient in random porous media, *Mathematical Geology*, **33**, 63–93.

Adler, P.M. and Thovert, J.-F. (1999). *Fractures and fracture networks*. Kluwer, Dordrecht.

Ameen, M.S. (1995). *Fractography: fracture topography as a tool in fracture mechanics and stress analysis*. The Geological Society, London.

Balberg, I. Binenbaum, N. and Wagner, N. (1984). Percolation thresholds in the three-dimensional sticks system, *Phys. Rev. Lett.*, **52**, 1465–1468.

Barenblatt, G.I. and Zheltov, Yu.P. (1960). Fundamental equations of filtration of homogeneous liquids in fissured rocks, *Soviet Dokl. Akad. Nauk*, **132**, 545–548.

Barenblatt, G.I., Zheltov, Iu.P. and Kochina, I.N. (1960). Basic concepts in the theory of seepage of homogeneous liquids in fissured rocks, *Soviet Appl. Math. Mech. (P.M.M.)*, **24**, 852–864.

Barton, N., Bandis, S. and Bakhtar, K. (1985). Strength, deformation and conductivity coupling of rock joints, *Int. J. of Rock Mech. and Min. Sci. and Geomech. Abstr.*, **22**, 121–140.

Békri, S., Thovert, J.-F. and Adler, P.M. (1997). Dissolution and deposition in fractures, *Engineering Geology*, **48**, 283–308.

Berkowitz, B. and Adler, P.M. (1998). Stereological analysis of fracture network structure in geological formations, *J. Geophys. Res. B*, **103**, 15339–15360.

Bogdanov, I.I., Mourzenko, V.V., Thovert, J.-F. and Adler, P.M. (2003a). Effective permeability of fractured porous media in steady-state flow, *Water Resourc. Res.*, **39**, doi:10.1029/2001WR000756.

Bogdanov, I.I., Mourzenko, V.V., Thovert, J.-F. and Adler, P.M. (2003b). Two-phase flow through fractured porous media, *Phys. Rev. E*, **68**, 026703.

Bogdanov, I.I., Mourzenko, V.V., Thovert, J.-F. and Adler, P.M. (2003c). Pressure drawdown well tests in fractured porous media, *Water Resourc. Res.*, **39**, doi:10.1029/2000WR000080.

Bogdanov, I.I., Mourzenko, V.V., Thovert, J.-F. and Adler, P.M. (2007). Effective permeability of fractured porous media with power-law distribution of fracture sizes, *Phys. Rev. E*, **76**, 036309.

Bonnet, E., Bour, O., Odling, N.E., Davy, P., Main, I., Cowie, P. and Berkowitz, B. (2001). Scaling of fracture systems in geological media. *Reviews of Geophysics*, **39**, 347–383.

Bossart, P., Meier, P.M., Moeri, A., Trick, T. and Mayor, J.-C. (2002). Geological and hydraulic characterisation of the excavation disturbed zone in the Opalinus Clay of the Mont Terri Rock Laboratory, *Engineering Geology*, **66**, 19–38.

Boulton, N.S. and Streltsova, T.D. (1977). Unsteady flow to a pumped well in a fissured water-bearing formation, *J. Hydrology*, **35**, 257–270.

Broadbent, S.R. and Hammersley, J.M. (1957). Percolation processes, I. Crystals and mazes, *Proc. Camb. Phil. Soc.*, **53**, 629–641.

Brown, S.R., Kranz, R.L. and Bonner, B.P. (1986). Correlation between the surfaces of natural rock joints, *Geophys. Res. Letters*, **13**, 1430–1433.

Brown, S.R. (1989). Transport of fluid and electric current through a single fracture. *J. Geophys. Res. B*, **94**, 9429–9438.

Celia, M.A., Bouloutas, E.T. and Zarba, R.L. (1990). A general mass-conservative numerical solution for the unsaturated flow equation. *Water Resourc. Res.*, **26**, 1483–1496.

Charlaix, E., Guyon, E. and Rivier, N. (1984). A criterion for percolation threshold in a random array of plates, *Solid State Commun.*, **50**, 999–1002.

Darcy, H.P.G. (1856). *Les fontaines publiques de la ville de Dijon*. Dalmont, Paris.

Dietrich, P., Helmig, R., Sauter, M., Hötzl, H., Köngeter, J. and Teutsch, G. (Eds.) (2005). *Flow and transport in fractured porous media*. Springer, Berlin.

Earlougher, R.C. (1977). *Advances in Well Test Analysis*. SPE Monograph Series, Dallas.

Faybishenko, B., Witherspoon, P.A. and Gale, J. (2005). *Dynamics of Fluids And Transport in Fractured Rock*. Geophysical Monograph, AGU.

Fischer, M.E. (1971). The theory of critical point singularities, in Green M.S. (Ed.) *Critical phenomena*, Proc. 51st Fermi School, Varenna, Italy, Academic Press, New York.

Gentier, S. (1986). *Morphologie et comportement hydromécanique d'une fracture naturelle dans le granite sous contrainte normale*, Ph.D. Thesis, Univ. d'Orléans, France.

Gerbaux, O., Buyens, F., Mourzenko, V.V., Memponteil, A., Vabre, A., Thovert, J.-F. and Adler, P.M. (2010). Transport properties of real metallic foams, *J. Colloid Interf. Sci.*, **342**, 155–165.

Garcia-Gonzales, R., Monnerau, C., Thovert, J.-F., Adler, P.M. and Vignes-Adler, M. (1999). Conductivity of real foams *Colloids and Surfaces A*, **151**, 497–503.

Gonzalez-Garcia, R., Huseby, O., Thovert, J.-F., Ledésert, B. and Adler, P.M. (2000). Three-dimensional characterization of a fractured granite and transport properties *J. Geophys. Res. B*, **105**, 21,387–21,401.

Gradshteyn, I.S. and Ryzhik, I.M. (1965). *Tables of integrals, series and products*. Academic Press, New York.

Gupta, A. and Adler, P.M. (2006). Stereological analysis of fracture networks along cylindrical galleries, *Mathematical Geology*, **38**, 233–267.

Hamzehpour, H., Mourzenko, V.V., Thovert, J.-F. and Adler, P.M. (2009). Percolation and permeability of networks of heterogeneous fractures, *Phys. Rev. E*, **79**, 036302.

Hassanzadeh, H. and Pooladi-Darwish, M. (2006). Effects of Fracture Boundary Conditions on Matrix-fracture Transfer Shape Factor, *Transp. Porous Media*, **64**, 51–71.

Hestir, K. and Long, J.C.S. (1990). Analytical expressions for the permeability of random two-dimensional Poisson fracture networks based on regular lattice percolation and equivalent media theories, *J. Geophys. Res. B*, **95**, 21,565–21,581.

Huseby, O., Thovert, J.-F. and Adler, P.M. (1997). Geometry and topology of fracture systems, *J. Phys. A*, **30**, 1415–1444.

Isihara, A. (1950). Determination of molecular shape by osmotic measurement, *J. Chem. Phys.*, **18**, 1446–1449.

Kant, R. (1996). Statistics of approximately self-affine fractals: random corrugated surface and time series, *Phys. Rev. E*, **53**, 5749–5763.

Kirkpatrick, S. (1971). Classical transport in disordered media: scaling and effective medium theories, *Phys. Rev. Lett.*, **27**, 1722–1725.

Koudina, N., Gonzalez–Garcia, R., Thovert, J.-F. and Adler, P.M. (1998). Permeability of three-dimensional fracture networks, *Phys. Rev. E*, **57**, 4466–4479.

Ledésert, B., Dubois, J., Velde, B., Meunier, A., Genter, A. and Badri, A. (1993). Geometrical and fractal analysis of a three-dimensional hydrothermal vein network in a fractured granite, *J. Volc. Geotherm. Res.*, **56**, 267–280.

Madadi, M. and Sahimi, M. (2003). Lattice Boltzmann simulation of fluid flow in fracture networks with rough, self-affine surfaces, *Phys. Rev. E*, **67**, 026309.

Malevich, A.E., Mityushev, V. and Adler, P.M. (2006). Stokes flow through a channel with wavy walls, *Acta Mechanica*, **182**, 151–182.

Malinouskaya, I., Mourzenko, V.V., Thovert, J.-F. and Adler, P.M. (2008). Wave propagation through saturated porous media, *Phys. Rev. E*, **77**, 066302.

Mandelbrot, B.B. (1982). *The fractal geometry of nature*. W.H. Freeman, New York.

Mardia, K.V. (1972). *Statistics of directional data*. Academic Press, London.

Marino, S., Coelho, D., Békri, S. and Adler, P.M. (2000). Electroosmotic phenomena in fractures *J. Colloid Interf. Sci.*, **223**, 292–304.

Matthai, S.K., Mezentsev, A. and Belayneh, M. (2005). Control-volume finite-element two-phase flow experiments with fractured rock represented by unstructured 3D hybrid meshes, in SPE Reservoir Simulation Symposium, SPE93341, 31 January–2 February, SPE, Houston, Texas.

Méheust, Y. and Schmittbuhl, J. (2001). Geometrical heterogeneities and permeability anisotropy of rough fractures, *J. Geophys. Res B*, **106**, 2089–2102.

Mourzenko, V.V., Thovert, J.-F. and Adler, P.M. (1995). Permeability of a single fracture; validity of the Reynolds equation, *J. Phys. II France*, **5**, 465–482.

Mourzenko, V.V., Thovert, J.-F. and Adler, P.M. (1996a). Geometry of simulated fractures, *Phys. Rev. E*, **53**, 5606–5626.

Mourzenko, V.V., Békri, S., Thovert, J.-F. and Adler, P.M. (1996b). Deposition in fractures. *Chem. Eng. Com.*, **148–150**, 431–464.

Mourzenko, V.V., Galamay, O., Thovert, J.-F. and Adler, P.M. (1997). Fracture deformation and influence on permeability *Phys. Rev. E*, **56**, 3167–3184.

Mourzenko, V.V., Thovert, J.-F. and Adler, P.M. (1999). Percolation and conductivity of self-affine fractures, *Phys. Rev. E*, **E59**, 4265–4284.

Mourzenko, V.V., Thovert, J.-F. and Adler, P.M. (2001). Permeability of self-affine fractures, *Transp. Porous Media*, **45**, 89–103.

Mourzenko V.V., Yousefian, F., Kolbah, B., Thovert, J.-F. and Adler, P.M. (2002). Solute transport at fracture intersections, *Water Resourc. Res.*, **38**, doi:10.1029/2000WR000211.

Mourzenko, V.V., Thovert, J.-F. and Adler, P.M. (2004). Macroscopic permeability of three-dimensional fracture networks with power-law size distribution, *Phys. Rev. E*, **69**, 066307.

Mourzenko, V.V., Thovert, J.-F. and Adler, P.M. (2005). Percolation of three-dimensional fracture networks with power-law size distribution, *Phys. Rev. E*, **72**, 036103.

Mourzenko, V.V., Thovert, J.-F. and Adler, P.M. (2009). Macroscopic properties of polydisperse, anisotropic and/or heterogeneous fracture networks, Proceedings of the International Conference on Rock Joints and Jointed Rock Masses, Tucson, Arizona, USA, January 7-8.

Mourzenko, V.V., Thovert, J.-F. and Adler, P.M. (2011a). Trace analysis for fracture networks with anisotropic orientations and heterogeneous distributions, *Phys. Rev. E*, **83**, 031104.

Mourzenko, V.V., Thovert, J.-F. and Adler, P.M. (2011b). Permeability of isotropic and anisotropic fracture networks, from the percolation threshold to very large densities, *Phys. Rev. E*, **84**, 036307.

Mourzenko, V.V., Bogdanov, I.I., Thovert, J.-F. and Adler, P.M. (2011c). Three-dimensional numerical simulation of single-phase transient compressible flows and well-tests in fractured formations, *Mathematics and Computers in Simulation*, **81**, 2270–2281.

Mualem (1976). A new model for predicting the hydraulic conductivity of unsaturated porous media, *Water Resourc. Res.*, **12**, 513–522.

National Research Council (U.S.) (1996). *Committee on Fracture Characterization and Fluid Flow, Rock fractures and fluid flow: contemporary understanding and applications.* National Academies Press.

Nelson, R.A. (2001). *Geologic analysis of naturally fractured reservoirs.* Gulf Professional Publishing, Boston, 2nd edition.

Paluszny, A., Matthai, S.K. and Hohmeyer, M. (2007). Hybrid finite element-finite volume discretization of complex geologic structures and new simulation workflow demonstrated on fractured rocks, *Geofluids*, **7**, 186–208.

Papoulis, A. (1991). *Probability, Random Variables, and Stochastic Processes.* McGraw-Hill, New York, 3rd edition.

Patriarche, D., Pili, E., Adler, P.M. and Thovert, J.F. (2007). Stereological analysis of fractures in the Roselend tunnel and permeability determination, *Water Resourc. Res.*, **43**, doi:10.1029/2006WR005471.

Sahimi, M. (2000). Fractal-wavelet neural-network approach to characterization and upscaling of fractured reservoirs. *Computers & Geosciences*, **26**, 877–905.

Sahimi, M. (2011). *Flow and Transport in Porous Media and Fractured Rock: From Classical Methods to Modern Approaches.* Springer, Berlin.

Singhal, B.B.S. and Gupta, R.P. (2010). *Applied Hydrogeology of Fractured Rocks.* Springer, Berlin.

Sinha, S.K., Sirota, E.B., Garoff, S. and Stanley, H.B. (1988). X-ray and neutron scattering from rough surfaces, *Phys. Rev. B*, **38**, 2297–2311.

Sisavath, S., Mourzenko, V.V., Genthon, P., Thovert, J.-F. and Adler, P.M. (2004). Geometry, percolation and transport properties of fracture networks derived from line data, *Geophys. J. Int.*, **157**, 917–934.

Snow, D.T. (1969). Anisotropic permeability of fractured media, *Water Resourc. Res.*, **5**, 1273–1289.

Stauffer, D. and Aharony, A. (1994). *Introduction to percolation theory.* Taylor and Francis, London.

Thovert, J.-F. and Adler, P.M. (2004). Trace analysis for fracture networks of any convex shape. *Geophys. Res. Letters*, **31**, L22502.

Thovert, J.-F., Mourzenko, V.V., Adler, P.M., Nussbaum, C. and Pinettes, P. (2011). Faults and fractures in the Gallery 04 of the Mont Terri rock laboratory: Characterization, simulation and application, *Engineering Geology*, **117**, 39–51.

Turcotte, D.L. (1992). *Fractals and chaos in geology and geophysics*, Cambridge University Press, Cambridge.

van Genuchten, M.T. (1980). A closed form equation for predicting the hydraulic conductivity of unsaturated soils, *Soil. Sci. Soc. Am. J.*, **44**, 892–898.

Vignes-Adler, M., Le Page, A. and Adler, P.M. (1991). Fractal analysis of fracturing in two African regions, from satellite imagery to ground scale, *Tectonophysics*, **196**, 69–86.

Volik, S., Mourzenko, V.V., Thovert, J.-F. and Adler, P.M. (1997). Thermal conductivity of a single fracture, *Transp. Porous Media*, **27**, 305–326.

Warren, J.R. and Root, P.J. (1963). The behavior of naturally fractured reservoirs, *Soc. Pet. Eng. J.*, **228**, 245–255.

Witherspoon, P.A., Wang, J.S.Y., Iwai, K. and Gale, J.E. (1980). Validity of cubic law for fluid flow in a deformable rock, *Water Resourc. Res.*, **16**, 1016–1024.

Wooding, R.A. (1960). Instability of a viscous fluid of variable density in a vertical Hele-Shaw cell, *J. Fluid Mech.*, **7**, 501–515.

Yaglom, A.M. (1957). Some classes of random fields in n-dimensional space, related to stationary random processes, *Theor. Probability Appl.*, **2**, 273–320.

Zimmerman, R.W., Kumar, S. and Bodvarsson, G.S. (1991). Lubrication theory analysis of the permeability of rough-walled fractures, *Int. J. of Rock Mech. and Min. Sci. and Geomech. Abstr.*, **28**, 325–331.

Zimmerman, R.W. and Bodvarsson, G.S. (1996). Hydraulic conductivity of rock fractures, *Transp. Porous Media*, **23**, 1–30.

Index

adherence condition, 67, 68, 78
advancing front method, 91, 113, 126, 146
aperture
  average, 14, 17, 28, 72
  hydraulic, 72, 84, 110
  local, 3, 10, 16, 61, 66, 126
  variance, 28
average
  conditional, 28
  spatial, 11, 27
  statistical, 11, 27

Baget watershed, 34, 39
Bessel function, 58
block, 30, 41, 46, 135
boundary conditions
  flow through a fracture, 66, 67
  flow through a fracture network, 87, 88
  flow through a fractured porous medium, 112
  solute diffusion, 74
  solute transport, 78, 157
  well test, 126
Brazilian test, 66
Brownian solute, 78

capillary fringe, 129
chords, 33, 37; see also trace
clogging, 9, 111, 116
cluster, 20, 22; see also component
component
  connected, 19, 44
  percolating, 19, 20, 22, 44, 55
  see also cluster
compressibility, 125
compressible flow, 8, 121, 125, 151
conditioning, 155
conductivity, 7, 21
  electrical, 74
conjugate gradient algorithm, 93, 116
consistency relations, 53
contact zones, 10, 16
continuity equation, 126, 130
continuum percolation, 1, 19, 43
control volume, 93, 116, 133
convection, 7, 78
convection-diffusion equation, 78
convection-dispersion equation, 79

convergence criterion, 93, 116, 155
convex objects, 42
correction factors, 58, 104
CPU time, 146, 149, 154, 155
crack, 9
critical probability, 20, 43
cubic law, 69, 72, 102
cut-off length, 13, 27

Damköhler number, 156
dangling ends, 98
darcy (unit), 5
Darcy equation, 6, 7, 87, 109, 110, 125, 153
decay length, 38, 60
density
  dimensionless, 7, 32, 43, 48, 50, 56, 59, 90, 148
  of fractures, 30, 37, 53, 146
  of intersections, 32, 49, 52, 58, 59, 126
  of solid blocks, 46
  of traces, 32, 50, 52, 58, 59, 61
diffusion, 7, 74, 75, 78
dip, 33
discretization
  effect, 94, 99
  of equations, 7, 92, 116, 133
  parameter, 91, 94, 99, 113, 114, 147, 150
dispersion, 7, 78
dispersion tensor, 79
displacement of fluid, 136
double porosity model, 125
drainage area, 127
drawdown test, 126

electrical conductivity, 74
electroosmotic coupling, 157
electroosmotic phenomena, 157
equilibrium, 131, 133
ergodicity, 27, 139
error function, 45
excavation damaged zone (EDZ), 152
excluded volume, 1, 6, 7, 41, 42, 48, 50, 53, 56, 58, 60
exponential law (fracture density), 38, 59

fault, 9
Fick's law, 74

finite volume method, 8, 92, 116, 130, 133
Fisher distribution, 38, 39, 58, 104
Fisher pole, 38, 39, 104
fissure, 9
flow experiment, 66, 87
foam, 152
Fourier transform, 16, 18, 26
fractals, 14, 22
fractional contact area, 17
fractional open area, 17, 24, 61, 80, 83, 84, 106
fracture, 2
  area, 42, 48
  conductivity, 7, 75–77, 82
  cross resistance, 110, 120
  extension, 30, 37, 86
  filled, 111, 116, 126
  flow rate, 68, 87, 110, 130
  Gaussian, 7, 12, 80
  heterogeneous, 39, 54, 61, 106
  hydraulic conductivity, 69
  in plane orientation, 33, 37, 58
  non planar, 30, 35, 86
  orientation, 33, 37
  perimeter, 43, 48
  position, 37
  Reynolds conductivity, 76, 77
  Reynolds transmissivity, 71, 73
  self-affine, 7, 28, 82
  shape, shape effects, 37, 45–47, 56, 90, 97, 121
  transmissivity, 68, 69, 72, 82, 84, 86, 87, 102, 110, 117, 122, 126
  velocity field, 98, 100, 117
fracture border, 99
fracture family, 3, 30, 34, 35, 39, 49, 50, 90
fracture network, 3, 86
  anisotropic, 7, 38, 39, 49, 54, 57, 59, 104, 105
  connectivity, 43
  deterministic models, 36
  finite fractures, 99, 105
  generation, 35, 37, 40
  geometrical properties, 7, 40
  heterogeneous (or non uniform), 37, 59
  hierarchical, 38, 39
  infinite fractures, 89, 98, 104, 113
  isotropic, 1, 53, 57, 94
  meshing, triangulation, 35, 91, 146, 156

fracture network (cont.)
  monodisperse, 1, 3, 40, 103
  percolation, 43, 54, 56, 59, 97, 103, 122, 157
  permeability, 7, 87, 88, 93, 94, 96, 102
  polydisperse, 3, 37, 54, 102, 121, 152
  random models, 37
  reconstruction, 35
  spatially periodic, 41, 93
  uniform, 1, 57
  volumetric area, 90, 113, 124
fractured porous medium, 3, 8, 109
  dense networks, 124
  dilute regime, 122
  permeability, 7, 111, 117, 118, 122, 153

generic cases, 40
grain packing, 152
gravity, 130

homogenization theory, 67, 75, 130
Hurst exponent, 14, 24, 25, 72, 83

I^2OUD or IIOUD, 42, 49, 58, 94, 109, 118, 141
implicit time formulation, 134
inertial forces, 5
IOUD, 42, 50, 53, 54

joint, 9

La Peyratte granite, 35
lag, 11
Laplace equation, 7, 74
Lattice Boltzmann algorithm, 84
line data, 48
lubrication approximation, 66, 70, 81

macroscopic properties, 1, 2
matrix
  compressibility, 125
  permeability, 109, 125, 130, 134
  porosity, 125, 130
mean field approximation, 106, 157
mechanical deformation, 157
memory requirements, 146, 154
meniscus, 132
meshing
  domain decomposition, 149
  fracture network, 91, 156
  fractures, 7, 146
  matrix, 8, 113, 148, 156
mineralization, 9
molecular diffusion, 74, 78
moments of solute distribution, 78

n-rectangle, 46
Newtonian fluid, 4, 5, 66, 87

observation line, 31, 48; see also scanline
Ohm's law, 21
outcrop, 30, 32, 50, 53, 63, 64

Péclet number, 79, 156
parallelization, 149, 151, 155
percolation, 7, 19
  bond, 20, 43
  correlation length, 21
  critical exponents, 21, 45, 97
  finite-size effects, 22, 45, 95, 118, 122
  probability, 23, 29, 44, 141
  site, 20, 21, 43
  threshold, 20, 23, 29, 43, 45, 54, 56, 59, 118
  universality, 21
perfect mixing, 157
periodicity, 18
permeability, 3–5, 7, 68, 93, 94
Picard's scheme, 134, 156
Poiseuille flow, 68, 69, 71, 79
Poisson process, 33
pore size, 3, 132
porous medium, 3
porous rock matrix, 109, 125, 130
potential, 133
power law, 1
  block density, 47
  fracture size distribution, 7, 37, 54, 102, 109, 121, 152
  meshing, 147
  spectral density, 26
  surface properties, 28
  transmissivity, 102, 122
pressure diffusivity, 125, 126
pressure drawdown well test, 125
probability density function, 11
pseudo steady flow, 127
pseudo-diffusion, 44

random field
  autocorrelation, correlation, 14, 15, 24, 61
  correlated, 14, 24
  covariance, 24, 25
  Gaussian, 24, 61
  generation, 14, 24, 61
  linear combination, 15, 16
  spectral density function, 24, 26
  uncorrelated, 15, 16, 24
reservoir engineering, 126, 152
Reynolds approximation, 7, 72, 73, 75, 76
Reynolds equation, 70; see also Reynolds approximation
Reynolds number, 5, 66, 67, 87
Richards equation, 134
Roselend tunnel, 40

scanline, 63; see also observation line
Schönhart polyhedra, 115, 149
seepage velocity, 5, 6, 11, 88, 90, 109, 112, 125, 130
self-similarity, 22
separation, 11, 28, 68, 80, 81
shape factor, 46, 53, 125, 127
sinusoidal channel, 70, 81
skin factor, 127
Snow equation, 7, 89, 90, 96, 101, 103, 104, 113
solute, 74
  deposition/dissolution, 156
  transport at fracture intersections, 157
spacings, 33
  average, 34, 49, 101, 128, 148
  variogram, 34
spatial periodicity, 18, 44; see also periodicity
stereology, 32, 51–54, 57, 59, 157, 158
Stokes equations, 7, 66, 87
storage coefficient, 127
streamtube routing, 157
strike, 32, 33
subvertical, 52, 151
sugar box model, 36, 135
surface data, 50
surface elements, 116, 133
surface tension, 132
surfaces
  anisotropic, 12, 14
  autocorrelation, correlation, 11, 62
  correlation length, 12, 14, 62, 72, 80, 106
  covariance, 27
  Fourier spectrum, 27
  Gaussian, 11, 12, 72
  generation, 27, 62
  height distribution, 11
  intercorrelation, 12, 72, 73
  isotropic, 11
  of fractures, 3, 10
  phase function, 17, 61
  roughness, 11, 16, 72, 80
  self-affine, 14, 25, 27, 72
  uncorrelated, 17

tetrahedra, 113, 133
thermal diffusivity, 74
trace
  analysis, 32, 50
  azimuth, 32
  definition, 30
  density, 32, 50, 52
  distribution, 33
  in cylindrical tunnels, 157
  intersections, 32, 52
  length, 33, 52, 61
  length distribution, 33, 35

map, 30, 35, 64
orientation, 32, 34
see also chords
transient flow, 127, 138, 151, 153
two-phase flow, 8, 129, 155
    capillary function, 131, 141, 142, 155
    capillary number, 133, 143
    capillary pressure, 129, 131, 133
    constitutive equations, 129, 131, 139, 155
    generalized Darcy's equations, 130, 132, 139
    hysteresis, 139
    interfacial tension, 131
    irreducible (or residual) saturation, 132, 139
    mobility, 136
    non-wetting fluid, 130, 133
    parallel plane fractures, 8, 134
    relative permeability, 129, 130, 132, 133, 135, 139, 141, 142, 144, 155
    relative transmissivity, 130, 132, 133
    saturation, 130, 136, 138–140
    steady regime, 139
    sugar box model, 8, 135
    transient flow, 138
    van Genuchten relations, 8, 131
    wetting fluid, 130, 132, 133

unit cell, 41, 44

variogram, 34
vein, 9, 30
volume elements, 116, 133
volumetric fracture area, 90

wavy walls, 81
well, 125, 135
well test, 125, 151
wetting, non wetting fluids, 130
Wiener–Khintchine theorem, 25, 26